Communicate Science Papers, Presentations, and Posters Effectively

Please visit our companion website, http://booksite.elsevier.com/9780128015001, for *Communicate Science Papers, Presentations, and Posters Effectively* which has further resources for readers.

Resources for Readers include:

- Instructions on how to use LaTeX
- Example presentations
- Additional questions and answers

Communicate Science Papers, Presentations, and Posters Effectively

Gregory S. Patience
Daria C. Boffito
Paul A. Patience

AMSTERDAM • BOSTON • HEIDELBERG • LONDON
NEW YORK • OXFORD • PARIS • SAN DIEGO
SAN FRANCISCO • SINGAPORE • SYDNEY • TOKYO

Academic Press is an imprint of Elsevier

Academic Press is an imprint of Elsevier
125 London Wall, London, EC2Y 5AS, UK
525 B Street, Suite 1800, San Diego, CA 92101-4495, USA
225 Wyman Street, Waltham, MA 02451, USA
The Boulevard, Langford Lane, Kidlington, Oxford OX5 1GB, UK

Cover Image Credit: We would like to thank the Ministry of Culture-Directorate General of
Antiquities/National Museum of Beirut for their permission for use of the cover photo of the
Ahiram sarcophagus.

Notices
Knowledge and best practice in this field are constantly changing. As new research and experience
broaden our understanding, changes in research methods, professional practices, or medical
treatment may become necessary.

Practitioners and researchers must always rely on their own experience and knowledge in
evaluating and using any information, methods, compounds, or experiments described herein. In
using such information or methods they should be mindful of their own safety and the safety of
others, including parties for whom they have a professional responsibility.

To the fullest extent of the law, neither the Publisher nor the authors, contributors, or editors,
assume any liability for any injury and/or damage to persons or property as a matter of products
liability, negligence or otherwise, or from any use or operation of any methods, products,
instructions, or ideas contained in the material herein.

British Library Cataloguing in Publication Data
A catalogue record for this book is available from the British Library

Library of Congress Cataloging-in-Publication Data
A catalog record for this book is available from the Library of Congress

ISBN: 978-0-12-801500-1

For information on all Academic Press publications
visit our website at http://store.elsevier.com/

Typeset by SPi Global, India

Working together
to grow libraries in
developing countries

www.elsevier.com • www.bookaid.org

Contents

Preface

Writing and publishing are as important as research: your work is incomplete until you publish it. If no one cites your work, it is as if you haven't done anything (Kamat, 2015). Writing well increases how often people cite your work (Calgano, 2012).

Books, articles, and courses teach us how to write better but we are not reading them. Many articles drown in the passive voice, rely on crutch/zero verbs (use, perform, make, do, carry out, conduct) and introduce sentences with phrases like "It is seen that". Nature and Science insist that authors favor the active voice. Most of us don't publish in either journal. To write well, verbs are active and vigorous text is precise and concise and sentences are straightforward. Avoid the seven cardinal sins of writing: hedging, signposting, redundancy, self-consciousness, narcissism, boosting, and periphrasis.

How well do you write papers and present your research? We crafted 20 questions that test your knowledge of the minutiae of communicating science effectively

- forming sentences,
- formatting graphs and tables,
- expressing data appropriately, and
- recognizing the impact of your work through bibliometric measures.

If you answer 14 of the questions correctly, you communicate well. We substantiate our answers from the top scientific journals and other books dedicated to communicating science.

QUESTIONS

1. *Publishing*: Web of Science™ (WoS) Core Collection (2014) and Google Scholar (2014) are two citation databases that index publications of all forms. At the end of 2014, WoS indexed 39 million documents, including scientific articles (23 million), proceedings papers (6 million), meeting abstracts (4 million), book reviews (2 million) and editorials, letters, reviews, and news (4 million). What percentage of the scientific articles (in WoS) have been cited at least once?
   (a) more than 90%,
   (b) between 80% and 90%,

(c) between 50% and 80%,

(d) less than 50%.

2. *Publishing*: More people have cited Axel Becke's article on density functional thermochemistry (Becke, 1993) than any other between 1989 and 2014. How many citations has WoS indexed for this article?

    (a) fewer than 5000,

    (b) between 5000 and 15 000,

    (c) between 15 000 and 45 000,

    (d) more than 45 000.

3. *Writing style*: Passive voice, active voice, agents, and patients: writing clearly means that you identify the patient, the verb, and the agent. Select the best sentence?

    (a) We measured the pressure periodically.

    (b) The pressure was measured periodically.

    (c) It was shown that the pressure varied periodically.

    (d) Pressure varied periodically.

    (e) The pressure measurement was performed at intervals.

4. *Writing style*: Fewer words can be more powerful than many words. The sentence below has 19 words and contains a hedge and an ionized verb. What is the minimum number of words to express the same information?

    It is interesting to remark that these results indicate that deactivation phenomena are not important below $500\,°C$ *for PdO/ZrO$_2$-Y.*

    (a) less than 6,

    (b) 6–10,

    (c) 10–14,

    (d) 14–19.

5. *Reporting data*: What do error bars in graphs or uncertainty in variables $(x \pm \Delta_x)$ represent? (Multiple answers.)

    (a) standard deviation, σ;

    (b) standard error of the mean, $\frac{\sigma}{n}$;

    (c) confidence interval, $t(\alpha, n-1)\frac{\sigma}{n}$;

    (d) instrument resolution;

    (e) maximum and minimum of a range of measurements.

6. *Reporting data*: Significant figures pollute literature data: don't carry more than your experiments warrant. Order the list below from most certain to most uncertain:

    (a) $310\,°C$,

    (b) $583.15\,K$,

    (c) $310\,°C$ with a relative uncertainty $\Delta_T = 1\%$,

    (d) $583.15\,K$ with a relative uncertainty $\Delta_T = 1\%$.

7. *Reporting data*: The International Bureau of Weights and Measures (Bureau International des Poids et Mesures, BIPM) maintains and updates SI writing conventions. All physical quantities can be expressed with the SI,

but the BIPM does accept nonstandard expressions (e.g., bar for pressure). Identify all acceptable SI expressions to describe silica content:
(a) The mass fraction of silica was 3%.
(b) The mass fraction of silica was 0.03.
(c) The silica content was $3\%_{wt}$.
(d) The silica content was 3% (w/w).
(e) The silica content was 3 wt%.

8. *Graphs*: Understanding typography—font types typefaces, character height, line weight—can help create an aesthetic graph. Identify all correct approximations for a 0.35 mm thick line.
(a) 0.014″,
(b) 1 pt,
(c) 1.3 px,
(d) 0.08 em,
(e) 0.35 pt.

9. *Graphs*: The ideal character height in a graphs ranges between 2.5% and 5% of the *y*-axis. It should be closer to 5% in presentations, posters, and graphical abstracts. What is the best Arial font size for a 50 mm graph (*y*-axis).
(a) 2 pt to 4 pt,
(b) 4 pt to 6 pt,
(c) 5 pt to 8 pt,
(d) 6 pt to 9 pt.

10. *Graphs*: Line weights in presentations are heavier than in journal articles. They also depend on frame dimensions—lines should be thicker for wider graphs. What line weight should frame a 70 mm graph for a paper (and poster)? What should the line weight be for a presentation or graphical abstract? (Multiple choices possible.)
(a) hairline, hairline;
(b) 3 pt, 3 pt;
(c) 0.75 pt, 1.4 pt;
(d) 0.25 mm, 0.5 mm;
(e) 0.25 mm, 0.25 mm.

11. *Tables*: Graphs are more effective than tables at communicating trends. Tables are better to report descriptive data, property data, and data with multiple response variables and factors. Select the best visual constructs to separate columns of data.
(a) vertical lines,
(b) shading,
(c) space,
(d) (a) and (b),
(e) (b) and (c),
(f) (a), (b), and (c).

12. *Papers*: Titles are to articles what poetry is to literature. They should be concise and precise. However, words in titles are also a source for Internet search engines. What should be the maximum number of words for a title?
    (a) 12,
    (b) 15,
    (c) 18,
    (d) 21,
    (e) 24.

13. *Papers*: We reference other work to give credit, to give a framework to our research, to educate readers, and to direct readers to published procedures and methods (to keep your paper short). Between 2011 and 2014, *Nature* published over 3200 articles and letters that referenced 113 000 articles. What is the average age of the references (the difference between the year *Nature* published the paper and the year the reference article was published)?
    (a) 4,
    (b) 6,
    (c) 8,
    (d) 10,
    (e) 12.

14. *Papers*: Talk to your readers and audience rather than lecture them. In the classic writing style, writers recognize that readers are competent and will recognize the truth as you lead them through your work. Soggy prose seeks "to argue for the truth" (Pinker, 2014). For each sentence, match one of the seven cardinal sins of writing on the left with the corresponding sentence on the right (multiple choices).

1	Hedging	a	These results are extremely significant statistically and compare favorably with validation studies (Hatzivassiloglou and McKeown, 1997).
2	Signposting	b	In general these results show that a system with zero-cost identities does not require centralized allocation of identities to encourage cooperation (Lai et al., 2003).
3	Redundant	c	Here we report our new results on the samarium-arsenide… (Ren et al., 2008)
4	Self-conscious	d	More recently researchers have attempted to quantify the effects of anxiety on foreign language learning (Horwitz et al., 1986).
5	Narcissism	e	Whether established pests are suitable for attempted eradication is extremely controversial (Myers et al., 1998).
6	Boosting	f	In this section we shall evaluate the rate of recombination for nonequilibrium conditions (Shockley and Read, 1952).

15. *Papers*: What are the minimum requirements for coauthorship?
    (i) Conceive and design the study (or parts of it).
    (ii) Collect and analyze data.
    (iii) Interpret data.
    (iv) Draft the article.
    (v) Revise parts of it.
    (vi) Approve the final version.
    (vii) Agree to be accountable for the results.
        (a) (ii), (iv), (vi), (vii);
        (b) (i), (ii), (iv);
        (c) (iii), (v);
        (d) (vi), (vii).

16. *Presentations*: There should be fewer slides in a presentation than the number of minutes allotted for it. For 15 min presentations, 10 to 12 slides is a good target. However, don't put 5 min worth of content into each slide. At most, how many images (or bullets) should a slide have?
    (a) 3,
    (b) 4,
    (c) 5,
    (d) 6,
    (e) 8.

17. *Presentations*: Some people speak too quickly, while others speak so slowly they put you to sleep (which may be a blessing). If you are speaking quickly, you have too much material and you will lose your audience: if you speak 100 words per minute, a 15 min presentation equals 1500 words, which is about half of the text in a paper. What is the optimal cadence of a presentation in words per minute?
    (a) less than 90,
    (b) 90–120,
    (c) 120–150,
    (d) 150–180.

18. *Posters*: How many figures and images should a poster have at minimum?
    (a) 2,
    (b) 3,
    (c) 4,
    (d) 5.

19. *Posters*: What is the minimum font size to read the poster title from 2 m (6.5 ft)?
    (a) 24 pt,
    (b) 36 pt,
    (c) 48 pt,
    (d) 60 pt,
    (e) 72 pt.

20. *Plagiarism*: Identify any cases of self-plagiarism:
    (a) You copy a graph from one of your articles to a book without permission from the copyright holder.
    (b) You copy several paragraphs from your thesis into an article.
    (c) You take data from a graph and reproduce it (not identically) in another figure.
    (d) You copy text from your patent to a scientific article.
21. *Bonus*: Ahiram's sarcophagus is inscribed with Phoenician text. All alphabets originate from Phoenician characters. Translate what we inscribed on the side of the lid on the front cover of the book (Figure 0.1).

𐤋𐤔𐤀𐤉𐤋𐤉𐤀 𐤀𐤉𐤕𐤟𐤋𐤆𐤟

FIGURE 0.1 We inscribed these Phoenicians letters on the side of the lid of King Ahiram's sarcophagus picture.

All books on scientific communication advocate writing clearly and concisely. However, most journal articles drown in the passive voice, and many rely on ridiculous phrases such as *It is seen that* to introduce an idea. Many researchers consider that anything other than the passive impersonal voice is unprofessional. Kirkman (1971) surveyed members of the Institution of Chemical Engineers to determine what kind of writing they preferred. Although he said the response "should destroy for ever the myth that most engineers prefer a heavy impersonal style," only 46% of the readers preferred the active voice. Is the active voice better than the passive voice for technical communication?

Writing conventions incite visceral reactions (arguments):

- *Data are* or *Data is*?
- Two spaces or one space after a period?
- Sentences can't start with *and*, *but*, or *because*.
- Infinitives should not be split.
- Don't use a contraction.
- Capitalize words following colons.
- (Unrolling toilet paper from the top or the bottom?)

Passive voice versus active voice in scientific literature is not a conflicting writing convention: write active sentences! They communicate more clearly and concisely, and are easier to understand.

We recommend ways of writing and presenting. Often we are assertive about these *rules*, but many journals require you to conform a writing conventions: The Chicago Manual of Style, AMA Manual of Style, The ACS Style Guide, and Scientific Style and Format: The CSE Manual for Authors, Editors, and

Publishers. Check these guides first before using a contraction or allowing data to be singular.

The first chapter is all about publishing and why it is important. We discuss impact factors for journals and *h*-indices for authors, and identify the countries that publish the most and the research fields that are growing and shrinking.

In the second chapter, we outline best writing practices, the active voice, and concise and plain language. We introduce some fundamental concepts of sentence structure. You will delete the phrase *It is seen that* (and its derivatives) entirely from your scientific repertoire of poor English after reading this chapter. Our premise is that expressing information with fewer words is better: sentences should not boost (e.g., *very, significantly, extremely*), hedge (*might be considered, is a distinct possibility, our results imply*), or mutilate/ionize/nominalize verbs to create noun phrases (*calcination was done, for the production of performed, made a measurement*). Writing in the active voice is difficult and requires practice. We reproduce texts from the scientific literature, demonstrate the problems with them, and suggest alternatives. At the end of each chapter we list exercises for practice and provide our solutions in Appendix A.

Language should be precise, and it is debatable how precise it needs to be, but data is rarely precise to more than three significant figures! We summarize statistics, uncertainty, and uncertainty propagation, and insist that you report your data correctly (precisely but no more precisely than your data merits).

We give explicit instructions to produce acceptable graphs and tables in Chapters 4 and 5. You will be an expert in creating graphs with the correct font size, line thickness height, and width after you read this chapter.

The following chapters on papers, presentations, and posters apply our guidelines to these three forums to communicate your results. We adopt the IMRAD format—introduction, methods, results, and discussion—and go through each of these sections for a standard scientific communication. We draw on the top journals for inspiration (and rules). Although we structure our discussion with the IMRAD format, we recommend you be more creative with your presentations and posters. Every second counts, and uninformative section titles waste time and space.

The consequences of plagiarism can be devastating (Chapter 9). If your name is on the paper, you are responsible for its content.

ACKNOWLEDGMENTS

The inspiration to write this book comes from graduate students who struggle writing their first paper, their second paper, and every paper after that. We started by compiling a glossary of overused phrases and expressions from students, and when the list became too large, we summarized it in a document that Elsevier posted on its website (Patience et al., 2013).

This document energized us to go further. We scoured the literature, and found many books that address writing English, or reporting data, or preparing scientific documents. Some of them also describe how to prepare graphs and tables. *Nature Methods* published a series on reporting data and the *Journal of Physical Chemical Letters* published 20 editorials on publishing papers. Who needs more? What researchers need to do more is to read these texts, but we feel that isn't enough: they need to practice, they need feedback, and they need explicit directives. This book complements the corpus of literature on scientific communication.

Many people contributed their artistic skills, and many more scrutinized our language and composition. In particular, Caterina Rocca wrote the Phoenician letters and contributed images, Jason Robert Tavares, Patrice Perreault, Federico Galli, Danilo Klvana, Cristian Trevisanut, Jitka Kirchnerova, Roland Malhamé, Clara Santanto, Joceyln Doucet, Pierre Sauriol, and Frank Ajersch all read a chapter or two and made helpful suggestions, Nicolas Patience compiled much of the data in the Appendices, and Christian Patience and Brendan Patience collaborated.

ANSWERS

Here are our answers to the quiz.

1. *Publishing*: (b). On average, 86% of all articles have been cited at least once. In 2014, 82% of the papers written in the early 1980s had been cited. Citation rates are increasing: in 2014, 90% of the articles written in 1996 had been cited at least once. The WoS *All Databases* has 92 million articles as of December 2015. (You can retrieve them by typing *":"* in the Basic Search window with *Topic* category.)

2. *Publishing*: (d). WoS registered over 47 000 citations as of December 2014. Google Scholar indexed 22% more citations of this paper. Lowry et al. (1951) wrote the most cited paper in 1951 (cited 308 000 times up until 2014).

3. *Writing style*: (d). This is the best sentence, but it can be better.

   (a) We measured the pressure periodically.
   It doesn't matter who measured the pressure: *We* is an unnecessary agent. A better agent would be the instrument (e.g., a pressure trans-ducer with a resolution of x mbar).

   (b) The pressure was measured periodically.
   The is a classic passive sentence and it is vague. Identify an agent and the frequency.

   (c) It was shown that the pressure varied periodically.
   It was shown that is superfluous. Removing this phrase doesn't change anything (except making it shorter).

(d) Pressure varied periodically.
This is the best sentence, but it is still vague. State the frequency.

(e) The pressure measurement was performed at intervals.
The word *measurement* makes your sentence longer unnecessarily. If you know what the pressure was, something measured it.

4. *Writing style* (a).
PdO/ZrO$_2$-Y is stable below 500 °C.

   - *It is interesting to remark* is an unnecessary phrase.
   - *not important* is better stated *unimportant*.
   - *these results indicate*: state the results.

5. *Reporting data*: (a-e). All of them can represent uncertainty. You must state explicitly which one them you are using. Error bars in *Nature Methods* graphs are predominantly standard deviation (a) or standard error (b) (Krzy-winski and Altman, 2013a). The JCGM (2008) states that (b) represents the *experimental standard deviation of the mean*, and that it is incorrect to refer it to as the *standard error of the mean*.

6. *Reporting data*: When the uncertainty is missing, we assume that it equals half the smallest significant figure. To convert degrees Celsius to kelvin we add 273.15. If the instruments don't record temperature to within ± 0.01 °C, then don't add 0.15 K when you convert degrees Celsius to kelvin. Adding 273.15 K to 310 °C and keeping five significant figures increases the certainty 1000 times.

			Rank
(b)	583.15 K	± 0.005 K	1
(c)	310 °C $\pm$ 1 %	± 3 K	2
(a)	310 °C	± 5 K	3
(d)	583.15 K $\pm$ 1 %	± 6 K	4

7. *Reporting data*: (b). A space between 3 and % is missing in (a).

8. *Graphs*: (a-d). The line thickness of graphs ranges from 0.3 mm to 0.6 mm.

9. *Graphs*: (c) and (d). One inch equals 72 pt. A two-inch plot (51 mm) equals 144 pt. So, 2.5% and 4% of 144 pt are 3.6 pt and 5.8 pt, respectively. However, Roman numerals (and uppercase letters) are shorter than the total font size: they represent approximately 75%, so the font has to be bigger to meet our recommendation. The minimum font size in a graph is 6 pt, and the maximum is the caption font size.

10. *Graphs*: (c) and (d). Graph line weights in papers and posters range from about 0.75 pt to 1 pt (0.25 mm to 0.35 mm). Axis line weights in papers and posters are 1% of the length of the *y*-axis (in millimeters). We recommend that the line weight in presentations and graphical abstracts be about double what they would be in papers and posters.

11. *Tables*: (c). We prefer white space to vertical lines or shading.

12. *Papers*: (b). The top 1000 most cited articles in *Science* and *Nature* (in 2013) average 10 words. Only 13 of the top 1000 papers have more than 15 words. Among the top 1000 most cited articles in WoS (until the end of 2014) 130 titles have more than 15 words. The longest title has 32 words. The shortest title has one word and six characters. The titles of TED talks average 5.5 words per title with a standard deviation of 2 (1978 titles up until May 2015 - http://www.openculture.com/2014/06/1756-ted-talks-listed-in-a-neat-spreadsheet.html).

13. *Papers*: (c). References in *Nature* follow a Weibull probability distribution rather than a normal distribution. The cumulative distribution function, $P(t)$, has a modulus $m = 1$ and a scaling parameter $\theta = 8.1$:

$$P(t) = 1 - \exp\left(-\frac{t}{\theta}\right)^m.$$

This means 63% of the references in *Nature* are less than 8.1 years old, and 50% are less than 5.1 years old.

14. *Papers*: Some of the sentences have more than one cardinal sin of writing.

1	Hedging	b	In general these results
		d	have attempted to quantify
		e	attempted eradication
2	Signposting	f	In this section we shall evaluate
		c	Here we report
3	Redundant	c	new results
4	Self-conscious	e	is extremely controversial
5	Narcissism	d	More recently researchers
6	Boosting	a	extremely significant
		c	extremely controversial

15. *Papers*: (a). Taking responsibility for parts of the work is the minimum requirement for authorship (Royal Society of Chemistry, 2014), but that is insufficient. Authors must participate in writing and in manipulating data or designing the experiments (Shewan and Coats, 2010).

   - conceived and designed the study **or** parts of it, **or** collected, analyzed, **or** interpreted data; and
   - drafted the article **or** revised important parts of it; and
   - approved the article's final version; and
   - agreed to be accountable for the results.

16. *Presentations*: (d). Six 8 cm × 8 cm squares fit into a standard slide. This is the absolute maximum. More than 2 slides with 6 regions with graphs and text is too much information. Consider TED talks where speakers limit most slide to one image, one idea per slide and the maximum is 4. Gallo (2014) recommends that the first 10 slides should have no more than 40 words.

17. *Presentations*: (c). We have recorded people speaking at several conferences. Anything less than 100 words per minute seemed slow. Presentations approaching 160 words per minute were fast. An manager from General Motors gave a 1 h presentation and spoke over 160 words per minute for 1 h. The presentation was excellent, but there was so much information that it was difficult to follow. If you speak quickly, take more pauses. Dananjaya J. Hettiarachchi won the Toastmasters International World Championship of Public Speaking and spoke at 120 words per minute. Some TED lecturers—Al Gore, Hans Rosling, Becky Blanton—speak at rates between 130 words per minute and 160 words per minute, whereas others speak at rates exceeding 180 words per minute (Dlugan, 2012). Gallo (2014, pp. 82) holds speaking faster than 160 wpm is best (the rate of a casual conversation). However, the audience in TED talks are different than at international conferences many of whom English is their second or third language.

18. *Poster*: (e). Regardless how far people are away from your poster, the title should never be smaller than 72 pt. If you would like senior professionals, researchers and professors to read your text, bigger is better. Remember that 72 pt equals 1 in. (25.4 mm). In an oral presentation, the minimum font size is about 20 pt. Most poster text should be at least 32 pt.

19. *Poster*: (d). We surveyed the opinions on posters of 65 doctoral students, postdoctoral fellows, and engineering and chemistry professors: 80% agreed that posters should have at least three figures. Graphs and images must dominate a poster not text. At least five images are necessary for a poster.

20. *Plagiarism*: Self-plagiarism doesn't exist. If you are the copyright holder, you can copy it. When you publish a paper, you assign the copyright to the publisher. It is no longer yours. If you copy it without citing the source or getting permission, you are plagiarizing and infringing the copyright.

21. *Hint*: Phoenician reads from right to left.

Chapter 1

Publishing Industry

Chapter Outline

Writing well is as much art as it is science. Captivating your audience such that they read your entire article is art. Expressing ideas precisely in a systematic sequence and a logical structure is…science. It takes practice, effort, and time to write well—any sentence can take what seems like hours. The immediate reward for all your toil and trouble is when an editor accepts your manuscript.

You have to communicate your research: if you are not writing, you are not working. Editors publish work that is original, important, clear, and relevant to their journal and reject papers that don't meet these criteria. *Science* (2014) rejects 93% of the manuscripts it receives. The rejection rate is increasing across all journals (Kamat and Schaltz, 2014).

The first hurdle in the publishing process is getting past editorial assistants, who check for English quality, topic suitability, figure format, plagiarism, and overall look. Only one in five submissions gets to the editor (Science, 2014). This book will help you get past editorial assistants. We anticipate that it will also help convince the editor, the second hurdle, to send your work for review. Editors check the abstract, introduction, conclusions, and pertinence of the references. The letter to the editor describes why the work is important, how it is original, and why you are sending it to the journal you have selected (Soares and Thomas, 2014). If you don't address these points, your chances that the editor will take the time to send the manuscript to reviewers are slim.

The reviewers are the final hurdle: They examine the scientific content and consider how well you wrote the paper. If they can understand it, they can assess its significance and will be more inclined to accept it with *minor revisions*. Don't despair if they reject it. Address the reviewers' and editors' comments. Spend more time improving the paper and resubmit it to another journal in the same journal community. When the paper finally gets accepted, it will likely be cited more (Calcagno et al., 2012).

The end of the grueling publishing process is the satisfaction of seeing your work in a journal. Equally gratifying is to know that other people read it and cite it.

Here we present bibliometric indicators that gauge the extent of your contribution and prestige. We discuss the *h*-index for individuals and journals, as well as the IF. We introduce the $\phi_{\tau,\xi}$ factor to assign a rank to papers, to individuals, and even to scientific categories.

1.1 BIBLIOMETRICS

Bibliometric indicators—how many articles you publish, how many people cite these articles, and the journals that publish your work—are important for academic careers. Institutions rely on them when they hire and promote professors and research staff. These metrics help universities, governments, and corporations allocate funding and identify emerging and desirable research disciplines (Vieira and Gomes, 2010). Furthermore, they gauge the productivity of individuals, they quantify the impact of a university or a department, and they provide input for ranking both individuals and universities or departments. Students apply to highly ranked universities first, so these institutes select from among the best candidates. Choosing the best students helps maintain the institutes' productivity, prestige, and reputation.

University enrolment continues to increase, along with the number of scientific journals and the number of papers they publish. In 1990, Web of Science™ (WoS) (2014) indexed 920 000 new papers; in 2000, it indexed 1 300 000 new papers (Figure 1.1). In the following decade, the publication rate increased by 80 000 new articles per year, and in 2009 WoS indexed 2 million documents for the first time. It has more than 39 million documents dating from the beginning of 1989 to the end of 2014. The publication rate is increasing all over the world, and China has made the most progress.

The most important bibliometric indicators are the number of papers and the number of citations per paper. How do these translate to a researcher's prestige, impact, or success? After 5 years, if you publish 15 articles and 500 people cite your work, are you successful? The advantage of bibliometric indicators is that they are concrete numbers. They are open to public scrutiny and are relatively unbiased. The weakness of bibliometric indicators is that they are thus far inadequate to compare the performance of individuals across disciplines. Furthermore, the number of papers and the number of citations might not always correlate with the importance or the impact of the contribution. However, the advantages of assessing performance with indicators outweigh the drawbacks.

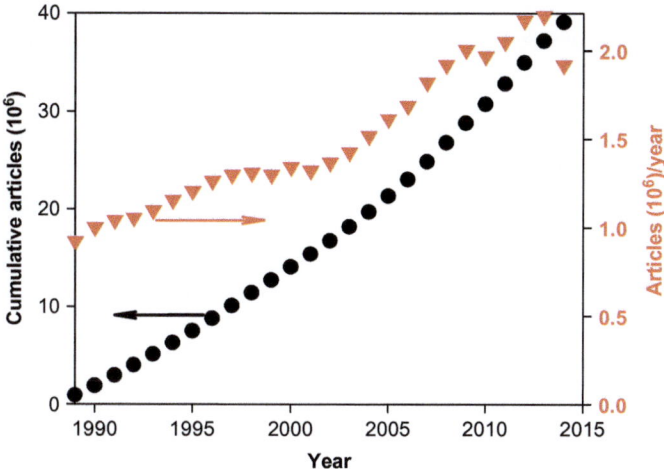

FIGURE 1.1 The number of articles appearing in WoS has been generally increasing year after year since 1989. In 2014, the number of new papers dropped by 277 000 versus 2013.

TABLE 1.1 Bibliometric Parameters of Excellence

	Percentile	IF	Citations (Φ)	Individual (Papers/ year)	Discipline (h_5-Index)	Country (Papers/ 1000/ year)
Outstanding	<0.1 to 1	30	750	12	300	3
Excellent	<10	4	60	6	150	2
Very good	<25	2.5	25	3	100	0.5
Good	<50	1.3	10	1.5	66	0.1

We classify performance as good, very good, excellent, and outstanding, which corresponds to a classification in the top 50%, 25%, 10%, and 1%, respectively (Table 1.1). Of the papers published 20 years ago, 50% have at least 10 citations ($\phi_{1,50}(1994) = 10$), 10% have at least 60 citations ($\phi_{1,10}(1994) = 60$), and 0.1% have at least 750 citations ($\phi_{1,0.1}(1994) = 750$). Several factors influence how many people read and cite papers. Articles in journals with high impact factors (IFs) are, by definition, cited more than those with low IFs. *Science* and *Nature* are high IF journals (over 30). The top 10% of journals have IFs over 4. The most prolific writers (and citers) are in the medical sciences and physics, whereas the least prolific are in the social sciences and mathematics. The top disciplines have at least 300 papers with 300 citations (h_5-index of 300); half of the scientific categories have at least 66 articles with at least 66 citations ($h_5 = 66$).

The number of publications per year is a controversial metric because it varies with the discipline. In fact, the IF and h_5 also vary according to the discipline. Moreover, senior individuals who have established a large research team produce more papers per year. Ioannidis et al. (2014) estimated that 15 million distinct authors contributed papers from 1996 to 2011. The Scopus database indexed 26 million documents for these researchers. Only 1% of the contributors published every year, but they accounted for 42% of all the papers: The top 1% produce 11 papers per year. Assuming four individuals coauthor the papers, the publication rate is 0.4 papers per person per year. Gingras et al. (2008) estimated that professors publish an average of three papers per year, which we arbitrarily classify as very good performance. We consider good performance to be 1.5 papers per year (Table 1.1).

Country Scientific Productivity

In the early 1990s, China was the 15th most prolific publisher of papers in the world (Table 1.2). The top 10 countries contributed 77% of all articles from 1989 to 1993, and doubled their output by 2010. The rest of the world quadrupled its productivity in the same period, and accounted for one-third of the publications between 2010 and 2014. China's contributions to scientific output was spectacular in this period, and increased 29-fold to 1.36 million papers, vaulting China to second place in the rankings. China now contributes half as many papers as the USA, whereas in the 1990s, England (the second-ranked country) published one-fifth as many articles as the USA.

The number of papers per capita loosely correlates with the standard of living. Among the top 15 nations, the Netherlands and Australia have the highest per capita publication rate, at 2.9 papers per 1000 individuals per year from the beginning of 2010 to the end of 2014. Canada and England contribute 2.4 papers per 1000 individuals per year, while the USA publishes at a rate of 1.8 papers per 1000 individuals per year. Several countries with a smaller population base have higher per capita publication rates, including Switzerland (4.3), Scotland (3.7), Iceland (3.7), Denmark (3.4), and Sweden (3.1) (Appendix B). The most industrialized nations publish at least one paper per 1000 individuals. The world would publish 3.5 million more papers a year if it could average just 0.5 publications per 1000 individuals.

University Ranking

To reach a modest target of 0.5 publications per 1000 individuals per year requires developing countries to found more universities. American universities dominate the ranking of the top institutes. ARWU (2014) ranks 17

TABLE 1.2 The Progression of the Number of Papers the Top 15 Countries Published from 1989 to 1993 and from 2010 to 2014 (Web of Science™, 2014)

Rank 1989-1993	Country	No. of Papers (Total 3 480 208)	Rank 2010-2014	Country	No. of Papers (Total 7 850 008)	Papers/1000/year
1	USA	1675717	1	USA	2854727	1.8
2	England	329190	2	China (PR)	1362268	0.2
3	Japan	272776	3	England	678294	2.4
4	Germany	232506	4	Germany	676481	1.7
5	France	206349	5	Japan	532608	0.8
6	Canada	202260	6	France	460756	1.4
7	USSR	143294	7	Canada	425287	2.4
8	Italy	116677	8	Italy	408716	1.4
9	Australia	91412	9	Spain	353427	1.5
10	Netherlands	84002	10	Australia	334642	2.9
11	India	79843	11	India	319807	0.1
12	Spain	66452	12	South Korea	302105	1.2
13	Sweden	61810	13	Netherlands	241974	2.9
14	Switzerland	54527	14	Brazil	234589	0.2
15	China (PR)	47480	15	Taiwan	174884	1.5

American universities in the top 20, and considers the number of publications, Nobel prizes, papers appearing in *Nature* and *Science*, the number of highly cited researchers, and per capita publication rate. Harvard University is ranked first in most of the criteria, and American universities lead in all categories except in the number of publications. Harvard University, the University of Toronto, Imperial College London, the University of Tokyo, and the University of São Paulo are the top five in the number of publications category.

Nine of the top 10 engineering schools are American—MIT (1), Stanford University (2), the University of California, Berkeley (3), (4) University of Illinois at Urbana-Champaign, (5) The University of Texas at Austin, (6) Georgia Institute of Technology, (7) University of California, Santa Barbara, (8) Imperial College of Science, (9) University of Michigan-Ann Arbor, and (10) Carnegie Mellon University.—on the basis of four criteria: highly cited researchers, number of indexed papers, number of papers published in high-impact journals, and funding. Considering only the number of indexed papers and the number of papers in high-impact journals, Chinese and East Asian universities populate the top 20. Only the Georgia Institute of Technology (15) and the University of California, Berkeley (19) break into the top 20 for these criteria.

The publication rate of Asia and the developing world is increasing rapidly. They are publishing more and more, but have not yet achieved the quality of American universities, which have more highly cited researchers. In the next sections, we discuss the indexing databases and classifying journals, individuals, and research fields according to how many papers are published and how many times people cite these papers.

1.2 CITATION INDICES

Throughout the book, we rely most on WoS Core Collection (Web of Science^{TM}, 2014). Its historical record is good, and it has exceptional search capabilities, with more than 14 topic classes—document type, author, editor, publication name, digital object identifier (DOI), year published, etc. Each of these classes has many subdivisions. WoS defines 251 scientific disciplines and 39 document classes. Seven of the 39 classes have 98% of the papers: 59% are articles, 16% are proceedings papers, 10% are meeting abstracts, and 13% are book reviews, editorials, reviews, and letters.

The second most important source of bibliographic data we cite is Google Scholar (GS) (Google Scholar, 2014). Scopus (2014) and PubMed are two other bibliographic databases. Scopus has more journal titles (22 000 up to July, 2014) and indexes papers from 5000 publishers (González-Pereira et al., 2013). PubMed concentrates on biology and medical disciplines.

WoS has many possible search options—name, date, topic, journal, and combinations of these. To access the 39 million WoS documents from the beginning of 1989 to the end of 2014, select "Year Published" and in the "Basic

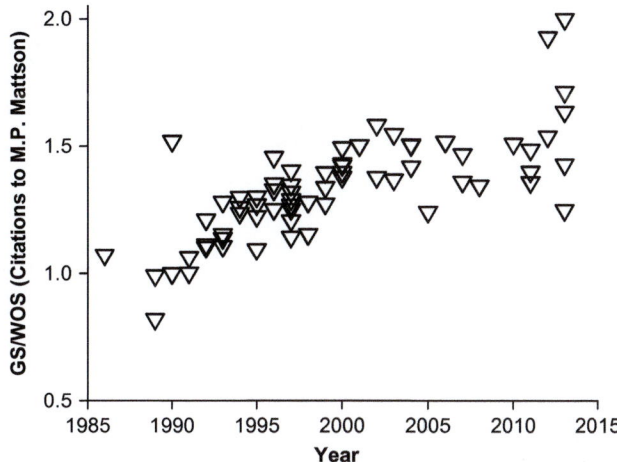

FIGURE 1.2 Historical citations of M.P. Mattson's papers. For the early 1990s, WoS indexes as many citations as GS. From 2000 on, GS indexes 40% more papers than WoS.

Search" field add 1989-2014. The "All databases" tool of WoS has 90.8 million articles in its archive from 1900 to 2014. Lowry et al. (1951) is the most highly cited paper in the "All database," "All years" search with 307 000 citations.

GS has as a larger database and includes more types of documents than WoS—symposia, patents, Chinese journals, theses, books, and proceedings. However, it indexes fewer papers from before about 1990. For example, it cataloged 192 000 citations for Lowry et al. (1951). To gauge the historical progression of the number of citations cataloged by WoS and GS, we examined the citation record of M.P. Mattson. GS registered 99 000 citations of his work (1281 documents) and WoS registered 66 000 (963 documents). For the late 1980s and early 1990s, WoS and GS index the same number of citations (Figure 1.2). From about 1993 to 2000, the ratio of the number of citations indexed by GS to that indexed by WoS increased steadily. From 2000, GS indexes about 40% more citations than WoS.

GS indexes 960 documents by M.P. Mattson that have at least one citation, whereas WoS has 780 documents. In June 2014, WoS indexed 1367 and 1169 citations of his top two articles, while GS indexed 1872 and 1763 citations. The citation profile of GS overlaps the WoS citation profile: a Tsallis power law distribution characterizes the number of citations as a function of paper rank (Figure 1.3). We assign 1300 to the ordinate of the number of citations indexed by WoS and 1800 to the ordinate of the number of citations indexed by GS.

GS cites 10% more documents in the chemical and materials sciences and engineering and computer sciences than does WoS. It cites 30% more in the health sciences, life sciences, physics, and mathematics. There are twice as many citations in GS as in WoS for the social sciences and humanities. Finally, for the

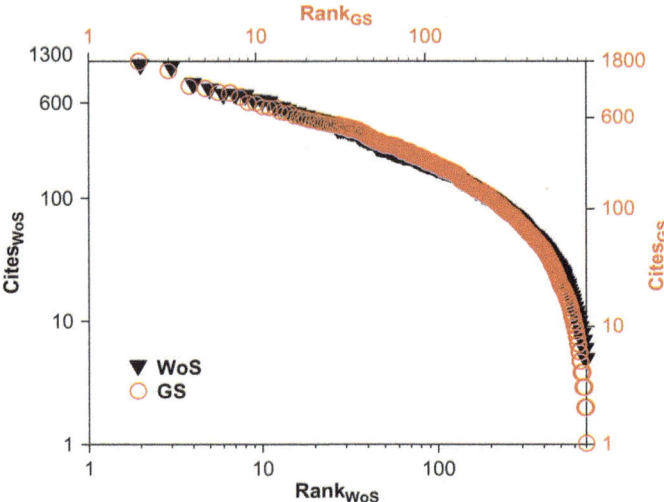

FIGURE 1.3   GS indexes more papers and more citations of Mattson's papers than WoS. The functional relationship between the number of papers and the number of citations is the same for both GS and WoS.

top three business journals, GS finds three times as many documents referencing these papers than does WoS. GS identifies 236 citations for the top paper in the *Journal of Business Ethics* (Barnea and Rubin, 2010), whereas WoS reports one-sixth of this number (42).

1.3   SCIENTIFIC CATEGORIES—DISCIPLINES

WoS indexes fewer recent documents than GS, but is recognized as the standard index database. Electrical and electronic engineering is the scientific discipline with the most papers—almost 2 million ($N = 1\,955\,254$). Slavic literature has the lowest number of papers: $N = 12\,400$. The number of papers published per year must correlate with the number of people working in the area and the amount of funding it attracts. Between 1989 and June 2014, WoS indexed an average of 780 000 papers per year ($\dot{N}_{1989\text{-}2014} = 780\,000$) for electrical and electronic engineering and an average of only 496 papers per year ($\dot{N}_{1989\text{-}2014} = 496$) for Slavic literature. A year-on-year increase in $\dot{N}$ indicates that the field is attracting interest (funding) and is growing. Between 2009 and 2014, WoS indexed an average of almost twice as many papers per year ($\dot{N}_{2009\text{-}2014} = 2\,200\,000$) compared with the average for the previous 20 years ($\dot{N}_{1989\text{-}2008} = 1\,250\,000$). The ratio of these two numbers is the ζ-index. An index greater than 2 indicates the field is growing faster than the average; an index close 1 means the field is stagnating.

Some fields have grown twice as fast as the average, whereas others have remained stagnant or even published less (Table 1.3). Many of the high-growth

TABLE 1.3 Growing Research Fields Based on the Ratio of the Average Yearly Output of Papers Published from 2009 to 2014 ($\dot{N}_{2009\text{-}2014}$) to That of Those Published from 1989 to 2008 ($\dot{N}_{1989\text{-}2008}$): $\zeta = \dfrac{\dot{N}_{2009\text{-}2014}}{\dot{N}_{1989\text{-}2008}}$

Rank	Category	$\dot{N}_{1989\text{-}2008}$	$\dot{N}_{2009\text{-}2014}$	h_5-Index	$\phi_{26,0.1}$	ζ
55	Energy fuels	10 095	30 764	143	238	3.0
67	Nanoscience and nanotechnology	6588	34 128	242	514	5.2
147	Nursing	3117	10 587	40	133	3.4
167	Mathematical and computational biology	2496	8023	119	796	3.2
181	Materials science and biomaterials	1804	6755	103	474	3.7
206	Agricultural engineering	1262	4214	78	300	3.3
224	Hospitality, leisure, sport, and tourism	795	3664	37	184	4.6
231	Integrative and complementary medicine	756	3345	41	188	4.4
235	Cell tissue engineering	513	3503	117	413	6.8
247	Cultural studies	331	1616	24	177	4.9

fields are related to biology and medicine. Cell tissue engineering has grown the most, followed by nanoscience and nanotechnology. The notable exceptions are cultural studies and hospitality, leisure, sport, and tourism. Energy fuels and nanoscience are the other two fields unrelated to biology that are growing proportionately more.

The slower-growing fields are mostly in the humanities—poetry, dance, and literary reviews. The last entry in Table 1.4, "Literature—countries," groups four categories together—(1) Slavic literature, (2) Romance literature, (3) German, Dutch, and Scandinavian literature, and (4) African, Australian, and Canadian literature. Surprisingly, the average number of papers published per year for aerospace engineering and marine engineering contracted substantially.

1.4 CITATIONS

Besides the number of papers, citations reflect the number of people working in a discipline. Citations relate to originality, relevance, and the interest of the scientific community (Gupta et al., 2005). The number of citations is a valid indicator within a discipline, but the citation culture differs across disciplines.

TABLE 1.4 Stagnating Research Fields Based on the Ratio of the Average Yearly Output of Papers Published from 2009 to 2014 ($\dot{N}_{2009\text{-}2014}$) to That of Published from 1989 to 2008 ($\dot{N}_{1989\text{-}2008}$): $\zeta = \frac{\dot{N}_{2009\text{-}2014}}{\dot{N}_{1989\text{-}2008}}$

Rank	Category	$\dot{N}_{1989\text{-}2008}$	$\dot{N}_{2009\text{-}2014}$	h_5-Index	$\phi_{26,0.1}$	ζ
97	Imaging science and photographic technology	8085	8141	70	291	1.01
116	Literature	6620	6365	16	27	0.96
121	Aerospace engineering	6566	4905	31	120	0.75
139	Music	4994	4830	18	37	0.97
143	Literary reviews	4880	4203	5	10	0.86
144	Art	4692	4612	13	23	0.98
148	Materials science characteriza-tion/testing	4556	4370	31	88	0.96
173	Computer science and cybernetics	3419	3295	49	204	0.96
196	Materials science, paper, and wood	2327	2211	36	105	0.95
197	Classics	2288	2320	7	28	1.01
222	Marine engineering	1500	1273	20	19	0.85
233	Agricultural economics policy	1147	1110	27	125	0.97
237	Poetry	1004	747	4	11	0.74
238	Dance	985	726	5	7	0.74
240	Literature—countries	3474	3102	5	17	0.89

The most highly cited papers are from the medical sciences, physics, biology, and chemistry. Researchers in the social sciences and mathematics cite each other's work less frequently.

Scientific Category ($\phi_{26,0.1}$)

We propose adopting the number of citations as an indicator, and introduce two factors Φ and $\phi_{\tau,\xi}$. The former represents the total number of citations (regardless of time). The latter corresponds to the number of citations of a paper over a defined time frame τ (in years) with respect to a percentile threshold value ξ. For example, from the beginning of 1989 until the end of 2014 ($\tau = 26$), WoS assigned 1 095 648 articles ($\Phi = 1\,095\,648$) to the applied physics category. Of these papers, 1096 ($\xi = 0.1\%$) are cited more than 394 times—$\phi_{26,0.1} = 394$. We can define any threshold value and any time frame from which to compare papers within a category (or even between categories). We can normalize citation rates with this factor and compare papers (and individuals) across scientific disciplines.

$\phi_{26,0.1}$ for the combination of all categories of WoS is 485: from 1989 to the end of 2014, 39 154 papers were cited 485 times or more. *Science* published 67 858 documents over the same period; the 679rd document was cited 1163 times: $\phi_{26,0.1} = 1163$. If we consider only the scientific articles (22 940), *Science*'s $\phi_{26,0.1} = 1803$: the 229th article was cited 1803 times.

$\phi_{26,0.1}$ is greater than 600 for the highest 20 scientific categories, including medicine, biology, probability and statistics, multidisciplinary physics, and three categories related to psychology. $\phi_{26,0.1}$ is less than 30 for the lowest 20 categories, which are populated with subjects related to the humanities. Coincidentally, many of these categories are barely maintaining their publication rate (Table 1.4). Keep in mind that GS indexes twice as many documents as WoS in the social sciences and humanities. These low citation rates are due, in part, to the lower coverage of WoS.

$\phi_{26,0.1}$ depends on the category, and is independent of the number of papers. WoS has indexed 18 engineering categories. To reach $\phi_{26,0.1}$ (the "outstanding" threshold) requires 350 citations for electrical engineering and 165 citations for mechanical engineering (Table 1.5). The lowest value of

TABLE 1.5 Total Number of Papers (N) Assigned to Each Engineering Discipline Between 1989 and 2014 (June) Together with Their Outstanding ϕ Factor Criterion ($\phi_{26,0.1}$)

Rank	Engineering Discipline	N	$\phi_{26,0.1}$
1	Electrical and electronic	1 955 254	268
28	Chemical	501 415	255
29	Mechanical	496 474	165
60	Metallurgical	342 657	226
62	Civil	327 520	157
63	Computer science and software	322 259	217
69	Biomedical	276 267	330
80	Multidisciplinary	244 364	192
90	Environmental	217 187	332
102	Manufacturing	188 079	126
119	Industrial	159 940	151
121	Aerospace	155 844	120
186	Petroleum	64 526	58
191	Geological	62 457	148
206	Agricultural	46 324	302
222	Marine	36 363	19
228	Ocean	33 254	64
235	Cell tissue	27 821	413

Note: The disciplines with the highest $\phi_{26,0.1}$ are more research oriented and potentially have greater growth rates.

TABLE 1.6 The $\phi_{26,0.1}$ Factor for Scientific Categories with Between 400 000 and 500 000 Published Papers (N)

Rank	Category	N	$\phi_{26,0.1}$
30	Computer science and information systems	488 954	238
31	Gastroenterology and hepatology	487 186	345
32	Biotechnology and applied microbiology	480 960	535
33	Genetic heredity	480 747	765
34	Plant science	479 038	504
35	Psychiatry	476 969	444
36	Hematology	469 345	416
37	Applied mathematics	460 387	225
38	Astronomy and astrophysics	454 408	578
39	Peripheral vascular disease	446 292	528
40	Experimental medicine research	443 250	658
41	Organic chemistry	442 677	298
42	Mathematics	422 156	151
43	Analytical chemistry	412 447	336
44	Instruments and instrumentation	404 983	200
45	Microbiology	402 376	502

$\phi_{26,0.1}$ is for marine engineering (19), but there are only 36 000 articles in this category. The well-established engineering disciplines that are more application oriented—mechanical, civil, manufacturing, and industrial engineering—have a lower $\phi_{26,0.1}$. Engineering disciplines with a heavy biological or medical component have a higher $\phi_{26,0.1}$—cell tissue engineering (413), environmental engineering (332), and biomedical engineering (330). Agricultural engineering is another discipline with relatively few papers but a high $\phi_{26,0.1}$ (302).

Fifteen categories that include chemistry, mathematics, medicine, computer science, and biology each have between 400 000 and 500 000 papers (Table 1.6). $\phi_{26,0.1}$ of hereditary genetics (highest—765) is four times greater than that of mathematics (lowest—151). Mathematics and instruments and instrumentation have the lowest $\phi_{26,0.1}$, followed by computer science and information systems, organic chemistry, and analytical chemistry with 238, 298, and 336. $\phi_{26,0.1}$ is greater for the medical and biological categories. Astronomy and astrophysics follows medicine in the ranking, which is consistent with the engineering classification. We list $\phi_{26,0.1}$ for all 251 categories in Appendix B.

1.5 JOURNAL PRESTIGE

The ϕ factor could rank journals, but the two commonest indicators to assess a journal's reputation are the IF and the Hirsch index (h-index). The IF of a given year equals the average number of citations of articles published in the two preceding years. *Science*'s 2012 IF was 31.2: papers *Science* published in 2010

TABLE 1.7 IF Versus SJR Rank

IF	Journal	SJR	Journal
102	*CA: A Cancer Journal for Clinicians*	39.4	*Reviews of Modern Physics*
53.3	*New England Journal of Medicine*	30.0	*Annual Review of Immunology*
52.7	*Annual Review of Immunology*	29.9	*CA: A Cancer Journal for Clinicians*

and 2011 averaged 31.2 citations in 2012. Journals may also report their 5-year IF (the average number of citations for all articles published during the previous 5 years) on their website. CiteFactor (2012) lists IFs and other bibliometric indicators for 8200 scientific journals.

The h-index gauges a researcher's reputation. It equals the number of articles, h, that have been cited h or more times. It combines productivity (number of papers) with impact (number of citations). GS classifies journals according to their h_5-index—the number of articles that have been cited that many times in the last 5 years. *Science* has published 311 articles that have been cited at least 311 times. González-Pereira et al. (2013) (SCImago (2014)) developed a metric that accounts for the number of citations and includes a weight factor representing the prestige of the journal citing the paper (SJR). Their database contains 20 000 journal titles. Other indicators include the cited half-life, the eigenfactor score, the immediacy index, and the total number of citations (over a given time period).

The ranks of the "top" journals change with the database and the bibliometric indicator. The SCImago (2014) ranking of the top three journals is similar to that of the IF ranking (Table 1.7). Only the *New England Journal of Medicine* is among the top three for IF, SJR, and h_5-index. The top three journals in terms of $h_{5,2014}$-index are *Nature* (355), the *New England Journal of Medicine* (329), and *Science* (311).

Impact Factor

The IF follows a power-law relationship with journal rank (Figure 1.4). The top 20 journals in 2012 had an IF over 20. The IF drops almost linearly at a rank of 3000. Half the journals have IFs of less than 1.3, and 1000 journals (25%) have an IF greater than 3.5.

The IF, like $\phi_{26,0.1}$, varies with the discipline. Pure sciences and medicine have the highest IFs, followed by engineering, economics, and social science and humanities.

Many disciplines' top journals have an IF below 3.5. To calibrate the prestige of a journal within a discipline, examine the IF of the top three or top four

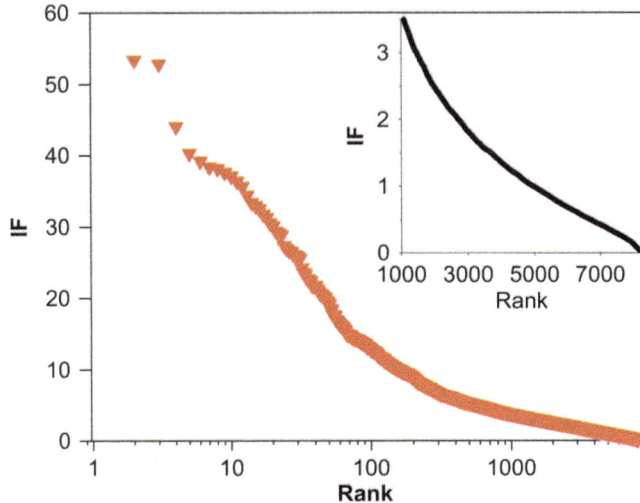

FIGURE 1.4 IF versus journal rank for 2012 (CiteFactor, 2012).

journals in that field. For example, the top journals in chemical engineering (based on 2014 IFs) are as follows:

Journal	IF
Chemical Engineering Journal	3.46
Chemical Engineering Science	2.43
AIChE Journal	2.26
Industrial & Engineering Chemistry Research	2.24

The *Chemical Engineering Journal* ranks 1108 among the 8000 journals (CiteFactor, 2012); the other three are ranked below 2000. All four journals are highly regarded in the field, but their IFs are low versus the IF of *Nature*, for example. When it comes to deciding where to publish (Kamat and Schaltz, 2013), the journal scope should guide the decision, not the IF. However, most studies can be tailored to fit the scope of many journals—topical journals or journals covering a general field—so the IF remains an important criterion.

h-Index

Journals report their IF on their website, but GS ranks journals on the basis of their h_5-index: the number of articles, h, that have been cited at least h times in the last 5 years. Between 2009 and June 2014, 100 journals had 107 articles that were cited 107 times or more ($h_{5,2014} = 107$) and *Nature* had the highest $h_{5,2014}$-index of 355. (The $h_{5,2013}$-index of WoS between 2008 and 2013 is 685 and its $h_{25,2014}$-index between 1989 and 2014 is 1921.)

TABLE 1.8 Rank of the Scientific Disciplines According to h_5-Index

Rank	Discipline	Journals in Top 100	h_5-Index (Top 3 Average) GS	WoS
1	Health and medical sciences	40	248	188
2	Physics and mathematics	19	153	118
3	Life science and earth sciences	16	311	248
4	Chemical and materials sciences	14	190	176
5	Engineering and computer sciences	7	173	156
6	Business, economics, and management	4	116	38
7	Social sciences	0	77	47
8	Humanities, literature, and arts	0	34	15

GS groups science and engineering into eight disciplines: health and medical sciences ranks as the top discipline according to the h_5-index, with 40 journals in the top 100, followed by physics and mathematics with 19 (Table 1.8). *Computers and Education* was the journal with the highest h_5-index (81) in the social sciences discipline. In the humanities, literature, and arts disciplines, the *Journal of Communication* was the highest, with $h_5 = 35$ (well out of the top 100). We averaged the h_5-indices of the three top journals in GS and selected the same journals to calculate their average h_5-indices from WoS. In all categories, the GS average is greater than the WoS average. The difference between the two indices is greatest for the social science, humanities, and business categories. GS indexes three times more citations than WoS in the business category. In the sciences, medicine, and engineering, GS indexes at most one-third more. It indexes 10% more in the chemical and materials sciences category.

The h-index is one of the best metrics to gauge individual performance, but it has several limitations (Bornmann and Daniel, 2007). It favors (1) well-established researchers (e.g., full professors), who are cited more often because of their reputation, (2) pure sciences, (3) individuals who collaborate in large teams, (4) hot areas of research, and (5) medicine and physics. It also depends on the bibliometric index—GS, WoS, or Scopus.

IF/SJR Versus *h*-Index

In 2014, twelve journals in the top 30 with the highest IF are published by Nature Publishing Group (Table 1.9). Five of the top 30 are published by Annual Reviews. Only three journals from Nature Publishing Group are in the top

TABLE 1.9 Citation Frequency in Top Journals (1989 to June 2014)

		Ranking According to h_5-Index					Ranking According to IF
Rank	h_5	Journal	ϕ (Rank)	IF (Rank)	Rank	IF	Journal
1	355	Nature	526 505 (1)	36 (11)	1	102	CA Cancer J. Clin.
2	329	New Engl. J. Med.	232 068 (8)	53.3 (2)	2	53.3	New Engl. J. Med.
3	311	Science	480 836 (3)	31.2 (18)	3	52.8	Annu. Rev. Immunol.
4	248	Lancet	158 906 (14)	38.3 (7)	4	43.9	Rev. Mod. Phys.
5	223	Cell	171 297 (12)	32.4 (16)	5	40.2	Chem. Rev.
6	217	Proc. Natl. Acad. Sci. USA	504 243 (2)	9.7 (162)	6	39.1	Nat. Rev. Mol. Cell Biol.
7	205	J. Clin. Oncol.	120 262 (23)	18.4 (53)	7	38.3	Lancet
8	193	Chem. Rev.	103 702 (29)	40.2 (5)	8	38.1	Nat. Rev. Genet.
9	191	Phys. Rev. Lett.	335 444 (6)	7.4 (254)	9	37.5	Nat. Rev. Cancer
10	190	J. Am. Chem. Soc.	408 307 (4)	9.9 (156)	10	37.0	Adv. Phys.
11	188	Nat. Genet.	76 456 (45)	35.5 (12)	11	36.3	Nature
12	181	JAMA	117 668 (25)	30.0 (20)	12	35.5	Nat. Genet.
13	178	Circulation	148 286 (17)	14.7 (68)	13	34.3	Annu. Rev. Biochem.
14	176	Chem. Soc. Rev.	35 918 (155)	28.8 (23)	14	33.3	Nat. Rev. Immunol.
15	174	Nano Lett.	75 287 (47)	13.2 (92)	15	32.8	Nat. Mater.

	N	Journal				Rank	IF	Journal
16	173	Adv. Mater.	79 860 (41)	13.9 (82)		16	32.4	Cell
17	171	Angew. Chem.	209 862 (9)	13.5 (91)		17	31.7	Energy Educ. Sci. Technol.
18	168	NBER Work. Pap.				18	31.2	Science
19	164	Nucleic Acids Res.	106 520 (26)	8.0 (218)		19	30.4	Nat. Rev. Neurosci.
20	162	J. Am. Coll. Cardiol.	70 298 (53)	14.2 (78)		20	30.0	JAMA
22	156	Blood	140 118 (18)	9.9 (157)		21	29.3	Nat. Photonics
23	153	Astrophys. J.	166 376 (13)	6.0 (368)		22	29.0	Nat. Rev. Drug Discov.
36	133	Cancer Res.	138 772 (19)	7.9 (227)		23	28.8	Chem. Soc. Rev.
47	128	Phys. Rev. B	278 680 (7)	3.7 (982)		24	27.3	Nat. Nanotechnol.
50	127	J. Neurosci.	150 228 (15)	7.1 (267)		25	26.9	Physiol. Rev.
70	117	Appl. Phys. Lett.	203 336 (10)	3.8 (892)		26	26.6	Cancer Cell
79	114	J. Biol. Chem.	402 449 (5)	4.8 (595)		27	26.5	Annu. Rev. Astron. Astrophys.
100	107	J. Immunol.	131 493 (20)	5.8 (394)		28	26.0	Nat. Immunol.
	73	J. Chem. Phys.	182 373 (11)	3.3 (1205)		29	26.0	Annu. Rev. Plant Biol.
	82	J. Geophys. Res.	149 781 (16)	3.0 (1438)		30	25.7	Annu. Rev. Neurosci.

30 of the h_5-index, and there are none from Annual Reviews. Of the top 20 SJR-ranked journals, 17 are in the top 30 journals with the highest IF; the SJR is an unnecessary index because it correlates with the IF ranking. Of the top 20 SJR journals, only 6 make it into the top 40 of the h_5-index.

GS ranks *Nature* as the top journal on the basis of the h_5-index. *CA: A Cancer Journal for Clinicians* doesn't break into GS's top 100 journals. Its $h_{5,2014}$-index is 59, which is about half that of the 100th-ranked journal of GS, but its IF is over 100—it's the top-IF ranked journal.

Rankings depend on how the indices weigh each bibliometric indicator—number of citations, number of documents, prestige of citing documents, and time frame. Both the IF and the SJR favor journals with few publications, whereas the h_5-index is generally highest for journals that publish thousands of articles per year ($N > 1000$). Twelve journals in the top 20 SJR ranking published fewer than 100 articles in 2012. Excluding *Nature*, which SJR classifies as 18th, the top 20 journals averaged 129 articles per year ($\dot{N} = 129$). Only one journal in the top 20 of GS's h_5-index classification published an average of fewer than 200 articles per year—the average was 1800 articles per year.

Nature published more than 24 398 articles from 1989 to 2014. *CA: A Cancer Journal for Clinicians* published only 1003 articles, reviews, editorials, and letters, and 70% of all its citations are of 25 papers on cancer statistics. These contributions are sui generis with respect to the other journals and artificially inflate its IF. An IF of 102 for this journal versus any of the top journals is unwarranted considering its narrow scope and the large number of citations of just 27 papers. The journals with the top three h_5-indices—*Nature*, the *New England Journal of Medicine*, and *Science*—publish a wider breadth of subjects and reach a larger audience; their impact is therefore greater.

1.6 ARTICLE CITATIONS

Scientific breakthroughs, new disciplines, and topics that impact multiple disciplines may become highly cited regardless of the journal. The *Journal of Physical Chemistry* has a modest IF (3.3), but it published the most highly cited paper from 1989 to 2014, that of Becke (1993), with 47 000 citations (Table 1.10). GS indexes 57 000 citations of this article. *Acta Crystallographica Section A* (IF of 3.1) published Sheldrick's 2008 article (Sheldrick, 2008), for which WoS logs 6400 citations per year and GS logs 8000 citations per year. Within a couple of years, it should surpass Becke (1993) in terms of number of citations per year. *Methods in Enzymology* has an IF of only 2.0, but it published the seventh most cited paper.

Nature is the only journal with an IF greater than 10 that published one of the top 10 papers. However, the average IF of the journals with the top 50 papers

TABLE 1.10 Top 50 Most Cited Articles (WoS 1989 to November 2014)

Rank	Journal	IF	Author	Year	Citations	Citations/year	References
1	J. Chem. Phys.	3.3	Becke	1993	46 444	2212	19
2	Nucleic Acids Res.	8.0	Thompson et al.	1994	40 473	2024	39
3	J. Mol. Biol.	4.0	Altschul et al.	1990	38 644	1610	22
4	Acta Crystallogr. A	3.1	Sheldrick	2008	38 382	6397	60
5	Nucleic Acids Res.	8.0	Altschul et al.	1997	36 608	2153	47
6	Phys. Rev. Lett.	7.4	Perdew et al.	1996	35 869	1993	121
7	Methods	4.0	Livak and Schmitgen	2001	29 372	2259	11
8	Methods Enzymol.	2.0	Otwinowski and Minor	1997	28 726	1690	40
9	Nucleic Acids Res.	8.0	Thompson et al.	1997	24 250	1426	27
10	Nature	36	Iijima	1991	23 023	1001	11
11	Phys. Rev. B	3.7	Kresse and Furthmüller	1996b	19 219	1068	76
12	Mol. Biol. Evol.	5.6	Tamura et al.	2007	18 468	2638	9
13	Acta Crystallogr. A	3.1	Sheldrick	1990	17 751	740	33
14	J. R. Stat. Soc. B	3.6	Benjamini and Hochberg	1995	16 160	851	14
15	Science	31	Novoselov et al.	2004	15 270	1527	16
16	Acta Crystallogr. D	13	Brunger et al.	1998	14 920	933	61
17	J. Chem. Phys.	3.3	Dunning	1989	14 702	588	54
18	J. Appl. Crystallogr.	5.2	Laskowski et al.	1993	14 510	691	22
19	Medical Care	3.4	Ware and Sherbourne	1992	14 435	656	45
20	Phys. Rev. B	3.7	Kresse and Joubert	1999	14 242	949	51
21	Bioinformatics	5.5	Posada and Crandall	1998	14 156	885	12
22	Phys. Rev. B	3.7	Blochl	1994	13 510	676	71
23	J. Appl. Crystallogr.	5.2	Kraulis	1991	13 498	587	14
24	Pharmacol. Rev.	20	Moncada et al.	1991	13 288	578	403
25	Comput. Mater. Sci.	1.5	Kresse and Furthmüller	1996a	13 241	736	71
26	Nature	36	O'Regan and Gratzel	1991	12 981	564	19

Continued

TABLE 1.10 Top 50 Most Cited Articles (WoS 1989 to November 2014)—Cont'd

Rank	Journal	IF	Author	Year	Citations	Citations/year	References
27	Nucleic Acids Res.	8.0	Berman et al.	2000	12 831	917	36
28	Phys. Rev. B	3.7	Perdew and Wang	1992	12 817	583	28
29	Acta Crystallogr. A	3.1	Jones et al.	1991	12 656	550	32
30	Bioinformatics	5.5	Ronquist and Huelsenbeck	2003	12 413	1128	5
31	N. Engl. J. Med.	53	Ross	1999	12 179	812	148
32	Mol. Biol. Evol.	5.6	Tamura et al.	2011	12 075	4025	41
33	Cell	32	Hanahan and Weinberg	2000	11 981	856	102
34	J. Comput. Chem.	4.6	Schmidt et al.	1993	11 869	565	141
35	Bull. Am. Meteorol. Soc.	6.0	Kalnay et al.	1996	11 734	652	59
36	JAMA	30	Cleeman et al.	2001	11 606	893	0
37	Acta Crystallogr. D	13	Emsley and Cowtan	2004	11 461	1146	19
38	Phys. Rev. B	3.7	Vanderbilt	1990	11 398	475	17
39	Phys. Rev. B	3.7	Perdew et al.	1992	11 374	517	121
40	Nature	36	Kresge et al.	1992	11 135	506	28
41	Bioinformatics	5.5	Huelsenbeck and Ronquist	2001	11 097	854	16
42	Struct. Equ. Model.	4.7	Hu and Bentler	1999	11 045	736	56
43	Nat. Mater.	33	Geim and Novoselov	2007	10 953	1565	91
44	Nature	36	Lander et al.	2001	10 922	840	450
45	J. Mol. Graph. Model.	2.1	Humphrey et al.	1996	10 759	598	19
46	Phys. Rev. B	3.7	Kresse and Hafner	1993	10 489	499	23
47	J. Appl. Crystallogr.	5.2	Spek	2003	10 472	952	30
48	N. Engl. J. Med.	5.3	Shamoon et al.	1993	10 210	486	43
49	Evolution	5.1	Rice	1989	10 131	405	3
50	Nat. Genet.	36	Ashburner et al.	2000	9999	714	33

is 11.5, which is high: highly cited papers cite previously highly cited papers (Bornmann et al., 2014). Kamat and Schaltz (2014) acknowledge that highly cited papers are promoted in journal home pages and are more likely to be noticed by researchers in the area, who include them as background information in their own papers.

Classifying Articles—$\phi_{\tau,\xi}$

WoS has indexed 39 million documents from the last 26 years (1989-2014); 23 million are classified as articles, and 14% of the articles remain uncited. Few articles are cited thousands of times. How many are cited 10 times or 100 times? If an article is cited 15 times in 5 years, how many times will it be cited after 20 years? More recent articles are cited less, but how long will others continue to cite the work?

The IF gauges the prestige of journals, and the h-index correlates with the productivity of an individual. We proposed the $\phi_{\tau,\xi}$ factor to rank scientific categories. It can also gauge how well an article is doing.

Whereas we set $\tau = 26$ to account for the historical significance of a category or journal, for an individual paper, we consider $\tau = 1$. We maintain $\xi = 0.1\%$ to represent outstanding performance, and classify $\xi = 10\%$ as excellent, $\xi = 25\%$ as very good, and $\xi = 50\%$ as good. An excellent paper in any given year has as many citations as the top 10% of papers published that year—regardless of the journal.

The citation history of articles follows a sigmoidal distribution: the number of citations increases with time and then approaches an asymptote.

$$\phi_{1,\xi}(t) = \frac{\Phi(\xi)}{1 + \exp\left(\frac{t_0 - t}{\beta_0}\right)},$$

where t represents the number of years since the article was published, $\Phi(\xi)$ is the asymptotic value of $\phi_{1,\xi}$, and t_0 and β_0 are fitted parameters.

The asymptotic value for 50% of the articles in WoS, $\Phi(50)$, is 10 (Figure 1.5). Coincidentally, GS reports the number of articles that have been cited 10 times to characterize individual performance—the I_{10}-index. In about 10 years, $\phi_{1,50}$ reaches $\Phi(50)$. It takes more time to reach the asymptote for $\xi = 25\%$ and $\xi = 10\%$, and as long as 20 years for $\xi = 0.1\%$. No papers cited more than 9000 times have reached their asymptotic value.

$\Phi(\xi)$	ξ (%)	Performance
750	0.1	Outstanding
60	10	Excellent
25	25	Very good
10	50	Good

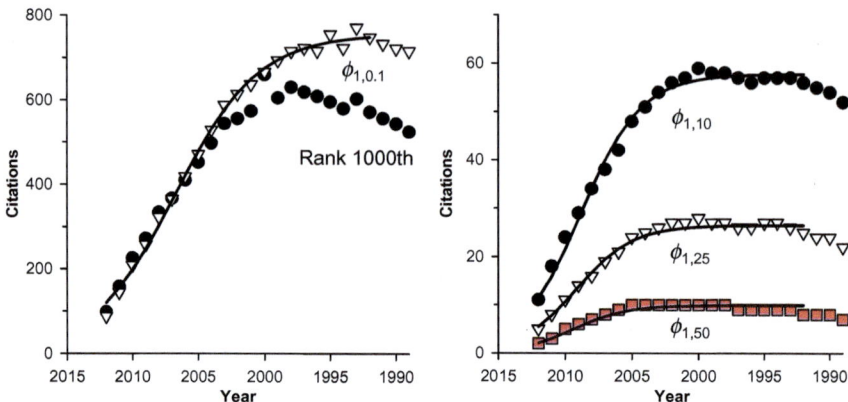

FIGURE 1.5   Citations as a function of time and $\phi_{1,\xi}$ (WoS).

After 5 years, good papers ($\xi = 50$) will be cited 5 times, very good papers will be cited 13 times, and excellent papers will be cited 30 times. Papers with $\xi = 0.1$ should expect to be cited 400 times after 5 years. To account for the dependence of $\phi_{1,\xi}$ on ξ, we multiply t by $\xi^{0.07}$.

Φ increases exponentially as ξ decreases; it is about 750 for $\xi = 0.1$ and 10 for $\xi = 50$ (Figure 1.5). These data follow a hyperbolic cosecant function:

$$\Phi(\xi) = \beta_1 \xi^{\beta_2} \, \mathrm{csch}(\beta_3 \xi).$$

This model accounts for 99.9% of the variance in the $\phi_{1,\xi}$ data from WoS. Equation 1.1 characterizes how many times people cite a paper with time:

$$\phi_{1,\xi}(t) = \frac{11.3\xi^{0.45} \, \mathrm{csch}(0.053\xi)}{1 + \exp\left(\frac{6.2 - t\xi^{0.07}}{2.7}\right)}. \tag{1.1}$$

We can apply it to estimate ξ for a given date and $\phi(t)$. For example, if a paper had 100 citations in 1992, what is its ξ? What percentage of the papers are cited more than 100 times? WoS registered 602 000 papers for 1992, of which 25 683 were cited 100 or more times, so $\xi = 4.3\%$. Solving for ξ in Equation 1.1 gives 3.9%, which underestimates the true value by 10%. The correlation agrees well with the data, so we conclude that we can accurately calculate the ranking, ξ, of any article in WoS without having to examine the citation index.

Increasing Citations

To maximize the visibility of your paper (i.e., to increase its number of citations), consider modifying the article features—coauthors, participating institutes, pages, references, and journal IF (Vieira and Gomes, 2010). For physics, mathematics, chemistry, biology, and biochemistry, the citation rate increases linearly with IF. Adding references enhances the number of citations by a factor of 2/3. Surprisingly, increasing the number of pages increases the number of citations by 70% for mathematics and biology/biochemistry and by 60% for chemistry and physics. Papers with more institutions and more authors are cited 50% more often for biology/biochemistry and mathematics, and 25% more often for chemistry and physics.

To increase a paper's visibility, and thereby citations, repeat words and phrases in the title, abstract, and keywords. Soares and Thomas (2014) recommend that keywords appear in the first 65 characters of the title. Further, they suggest searching for keywords in Google Trends and Google AdWords.

1.7 PERSONAL CITATION REPORTS

We said that

- you are not working if you don't write; furthermore,
- you are not working if you don't publish (Whitesides, 2004); but Kamat (2015) says
- you are not working if your papers aren't cited.

Journals compete to attract the best papers to improve their reputation and increase readership. The prestigious journals attract the best papers, and thereby maintain their stature. One of our objectives as researchers is to have the flexibility and freedom to innovate and create in scientific domains that interest us and impact humanity positively. Just like scientific journals, news media, and business, we have constraints, which include funding, space, and attracting highly qualified personnel. (This is more true for the pure sciences and engineering where financing a Ph.D. student requires 50 000 USD y^{-1} to 100 000 USD y^{-1})

Funding

Besides writing clearly and identifying how important your research is, your reputation weighs heavily on winning grant competitions. Funding agencies, universities, companies, and governments allocate their resources to individuals they consider most likely to succeed. Institutions judge researchers on the basis of their publishing track record (among other metrics), including how many papers (and patents) they have published, in what journal, the total number of citations, the average number of citations, and the h-index.

NSERC (2014) recognizes other forms of research contributions—books, monographs, conference/symposia proceedings, and government publications—and practical applications related to products, processes, designs, services, patents, and licenses.

Some funding agencies might count how many papers an academic has published in refereed journals in the preceding 5 years. These agencies reject grant proposals for individuals with fewer than 5-10 papers. They apportion the maximum weight to individuals with more than 30 papers. Publishing 6 papers a year ($\dot{N} = 6$) is excellent productivity, which gives 240 papers at the end of a 40-year career.

Ioannidis et al. (2014) estimated that 15 million authors published scientific papers between 1996 and 2011 (Scopus database). During this period, Scopus (2014) cataloged 25 million.

$\dot{N}$	Performance
12	Outstanding
6	Excellent
3	Very good
1.5	Average

Only 150 000 individuals published every single year from 1996 to 2011. These prolific writers accounted for 42% of the papers cited and $\dot{N} = 10.8$—outstanding performance. Besides publishing exciting research, publishing continuously correlates with reaching the largest audience.

To reach 30 papers in 5 years requires academics to collaborate or to maintain as many as three Ph.D. students, which represents about 300 000 USD y^{-1} or 60 000 USD per paper. Considering that the scientific community publishes 2 million articles a year, the cost of this research is $120 billion, which excludes patents and work in industry that remains unpublished. Operating costs and overhead in industry can range from $300 000 to $500 000 per principal investigator (including technicians). For the world to achieve a modest 0.5 papers per 1000 individuals per year (one-third of the current rate of the USA—Table 1.2), an additional yearly investment of $90 billion is required.

Metrics

Hirsch (2005) considered successful physicists as those who published 20 papers that were cited 20 or more times spanning 20 years ($h = 20$). Truly unique physicists achieved an h-index of 60.

h	Performance	Career Level
60	Unique	
40	Outstanding	
20	Excellent	Successful career
18	Very good	Tenure-track full professor
12	Good	Associate professor

The h-index depends on the field—unique economists might have an h-index of 40; unique linguists might have a lower h-index (especially if they publish most of their work in books). Harzing (2010) demonstrated that social science academics have as high an impact (citation rate) as academics in the sciences when one accounts for the number of authors per paper (h_i-index). The h_m-index corrects the h-index for age bias by dividing the h-index by the number of years since the individual first published.

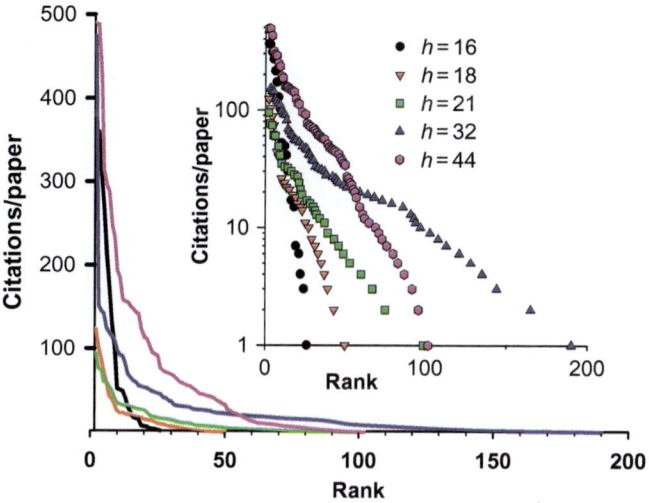

FIGURE 1.6   Author profile of citations per article versus rank and *h*-index.

The number of papers that you publish per year, the average number of citations per paper, your *h*-index, and the journals that publish your work are all indicators of productivity and success. The relationship between the number of citations per article versus rank, as for the IF and the number of journal citations per article, follows a power law: few publications have many citations (more than 60), and many have few citations (fewer than 10) (Figure 1.6). On a semi-log plot, $\log_{10}$ of the number of citations per paper drops almost linearly with the article rank. If the top five cited papers and those papers that have not been cited are excluded, an average of 50 citations per paper is excellent; an average 25 citations per paper is very good, and more than 12 citations per paper is good.

Individual *h*-indices increase with the number of publications, but some areas of research (and individuals) have high *h*-indices despite their relatively low number of publications: In Figure 1.6, the researcher with an *h*-index of 44 has published half as many articles as the individual with an *h*-index of 32.

Publishing in high-IF journals indicates success. Any journal with an IF greater than 3.5 is close to the top 10%—the excellent classification. About 50% of the journals have an IF greater than 1.3, which we classify as good.

1.8   EXERCISES

1. Most countries with lousy weather have high publication rates per capita. Name three southern countries with great weather and high publication rates.

2. How many papers does an average researcher publish each year?
3. How long does it take to write an article?
4. How is it possible to author 30 scientific papers in 1 year?
5. How many articles represent an excellent academic career? How many citations represent an excellent academic career?
6. What is the median IF in the Thomson Reuters Journal Citation Reports?
7. What is the median number of citations of an article?
8. Calculate your WoS and GS h-index and h_5-index.
9. A paper you wrote in 2009 has 15 citations as of 2014. How many citations do you expect it to have (for both WoS and GS) by 2024?
10. Which research teams have more than one paper in the 1989-2014 top 50?
11. How does the IF of *CA: A Cancer Journal for Clinicians* overstate its impact?
12. Why do journals that publish strictly review articles have higher IFs than journals in the same domain that publish original work?
13. Do journals based on reviews have higher h_5-indices?
14. What are the relative merits of the IF versus the h_5-index to rank journal prestige?
15. How do you compare the h_5-index of a mathematician with that of scientists and engineers?
16. Does a journal's IF continually increase? Do a researcher's h-index and h_5-index continually increase?
17. List five ways to increase the number of times your (future) article is cited.

Chapter 2

Writing Style

Chapter Outline

Good sentences are the foundation of good paragraphs and eventually good papers. Everyone agrees that sentences should be clear and understandable. But what does that mean? We assert that to write clearly means that the text be **concise** and **precise**, verbs be **active** and **vigorous**, and sentences be **straightforward**. These characteristics have some overlap so, to better categorize each, we include an antonym:

Concise	vs	**Redundant**	(words and text)
Precise	vs	**Vague**	(verbs and text)
Active	vs	**Passive**	(verbs)
Vigorous	vs	**Feeble**	(words and verbs)
Straightforward	vs	**Convoluted**	(sentences)

Concise can be precise, but in the context of this chapter we mean that it is nonrepetitive. Developing a *new innovation* is redundant since innovations are by definition *new*. Writing that something is *very hot* or *very fast* is imprecise and vague. We recommend replacing vague adjectives such as *very*, *significantly*, and *extremely* with numerical values. Identify the agent and let the verb express the agent's action on the patient. The agent is the person or thing that acts on the patient. Rather than *The surface area was determined by a porosimeter*, write *A porosimeter measured the surface area*. In the latter case, the porosimeter is the agent and the surface area is the patient. Ionizing a verb makes it feeble. *To determine* is better than *for the determination of*: leave the power in the verb, don't convert it to a noun. Finally, reformulate sentences with more than

Communicate Science Papers, Presentations, and Posters Effectively. http://dx.doi.org/10.1016/B978-0-12-801500-1.00002-4

40 words. Sentences are on average 15 words long (Scott, 1989). Sentences in the abstracts of *Nature* and *Science* are longer. In 2013, abstracts of the top 500 articles in *Nature* averaged 25 words per sentence, with a standard deviation of 10 words. *Science*'s top 500 articles averaged 22 words per sentence, with a standard deviation of 8. Only 5% of the sentences in *Science* have more than 36 words. Long sentences are often convoluted, making them difficult to understand.

We extract sentences from scientific articles, identify flaws, and recommend how to improve them. The left column contains sentences from the literature that are flawed. The right column demonstrates how to improve the sentence.

Redundant

We present a simple derivation of a simple generalized gradient approximation, in which all parameters (other than those in local spin density) are fundamental constants.
> Perdew et al. (1996, p. 3865)

Concise

We derive a generalized gradient approximation, in which all parameters (other than those in local spin density) are fundamental constants.

Vague

Coal pyrolysis to acetylene in hydrogen plasma is carried out under ultrahigh temperature and milliseconds residence time.
> Yan et al. (2013, p. 2119)

Precise

Hydrogen plasma pyrolyzes coal to acetylene above 1600 K in 5 ms.

Feeble

It should be acknowledged that this simple model does not appear to have universal applicability.
> Dudzik et al. (1998, p. 12661)

Vigorous

This model does not always apply.

Passive

Similar results were obtained when Sox2-Cre mice were injected in the same locations (Table S2).
> Friedmann-Morvinski et al. (2012, p. 1082)

Active

When we injected Sox2-Cre mice (with...) in the same places, the results were similar (Table S2).

Convoluted

Additionally, the OSIRIS measurements show the aerosol enhancement building most strongly above the Asian monsoon, which has recently been identified as a potentially effective pathway for transport of air from the troposphere to the stratosphere (5) and a potentially important region for the formation of small stratospheric aerosol particles during volcanically unperturbed conditions (6, 7).
> Bourassa et al. (2012, p. 78)

Straightforward

Asian monsoons carry air from the troposphere to the stratosphere (5), where OSIRIS measured higher aerosol concentrations; stratospheric aerosols may form in this region when volcanoes are inactive (6, 7).

It might take you longer to write a paper following these recommendations, but you will save your readers time. In fact, more people will read more of your paper since there will be less to read. The drawback is that you will begin to recognize poor grammar everywhere, which becomes irritating.

- Avoid uncommon words (Burger Associates Inc., 1991).
- Use plain language.
- Use contractions (Scott, 1989).
- Minimize problematic verbs (e.g., *perform* and *do*).

Use plain words you would say in a conversation. Scott (1989) speaks of a double vocabulary: simple words when we talk and complicated words when we write. Complicated words are harder to understand than their simpler counterparts.

The problematic verbs in science literature include *do*, *make*, *perform*, *try*, and *use* (Patience et al., 2013). Often these verbs replace verbs that convey an action: *to measure* is better than *to make a measurement*. When you replace *measure* with one of the poor substitutes, it takes more words to express an idea.

Verb	Phrase
Done	The experiments were done at an elevation of 2800 m
Tried	We tried low pressure to measure the equilibrium composition
Used	A gas chromatograph was used to measure the equilibrium composition
Performed	We performed experiments at an elevation of 2800 m
Made	We made measurements at an elevation of 2800 m with the use of a gas chromatograph

Logos, Ethos, Pathos

Written scientific communication maximizes **logos**—persuading with data and facts. **Ethos**—a mode of persuasion that relies on establishing yourself as an authority—is important in presentations (together with logos). **Pathos**—an emotional appeal—is common in fund raising, commercials and political speeches, but uncommon in scientific writing. Pathos dominated Bryan Stevenson's TED talk "We need to talk about an injustice" (Gallo, 2014) and he received the longest standing ovation to date. He spent 25% of his time with logos, 10% with ethos and 65% with pathos. Making an emotional connection with your audience is important to keep them engaged. However, be careful not to exaggerate emotional appeals or statistical evidence: an overwhelming consensus of scientists agree that climate change is like staring at the barrel of a loaded gun. Here, the **ethos** is *consensus of scientists* and the **pathos** is *the loaded gun*.

Let the data and facts speak. Both ethos and pathos require metadiscourse—verbiage about verbiage. The passive impersonal voice accompanies metadiscourse and this combination clutters (verbiage^{verbiage}).

Many classes of metadiscourse infect scientific writing. Three egregious ones are boosters, hedges, and signposting. Signposting is **redundant** in papers: Readers have the entire manuscript in front of them and can go back and forth between sections. Thoughtless signposting requires readers to put more effort into understanding the signs relative to what they point to (Pinker, 2014).

It should be noted that while we have already mentioned the interaction of the implants with the tissue and will discuss later in this manuscript the effect of the regenerative therapies on the tissues, the possibility exists that the materials delivered to the interface may interact with the implant itself. We will briefly discuss this possibility now before moving forward with the rest of our discussion.
Peramo and Marcelo (2010, p. 2018)

Materials at the interface can interact with the implant.

Good speakers use signposting in presentations: they begin with the conclusions, tell people how they got there, and restate them at the end of the talk. This ensures that people who get distracted at some point during the presentation can still understand the essence of the presentation.

Boosters and intensifiers are words and phrases that exaggerate certainty. They render sentences im**precise** and **vague**. Rather than state that something is *highly significant*, state its significance. Science doesn't allow its authors to use *significant* except to express statistical observations.

Hedges are another form of metadiscourse in which authors distance themselves from their own work: *It was found that*, *might*, *could*, and *possibility*. All measurements have uncertainty, and readers understand that. Hedging is necessary in adversarial legal documents and patents. If you feel your readers will "misinterpret a statistical tendency as an absolute law" (Pinker, 2014), qualify your generalizations. If your text is riddled with *would*, *suggest*, *appear*, and *seem*, then get better data or accept responsibility for what you did and delete these tiresome **feeble** words. Together with signposting, Peramo and Marcelo (2010) hedge: *the possibility exists that...may interact*.

2.1 CONCISE VERSUS REDUNDANT

Cleaning up redundancy is the easiest task when correcting a draft. First, though, you have to recognize it. Redundancy is pervasive—many researchers tend to add extra words, expressions, and sentences (Patience et al., 2013). We want to help you break this habit.

Empty Introductory Phrases

Useless introductory phrases, beginning with *It is…that* pollute scientific articles and theses. These phrases can take up to 30% of a sentence. Instead of saying *it is known that…*, state what is known. Who or what is *It*? Sometimes the construction is appropriate, but when adjectives such as *noticed* or *mentioned* follow, invariably it is extraneous:

- It is reported that…
- Thus it is of great concern that…
- It has to be noticed that…
- It should, however, be mentioned that…
- It is worth mentioning that…
- It is evident that (found that, assumed that)…
- It is important to use…
- Also, it is known that…
- It is suggested that…
- It was found that…
- Comparison of the results presented above reveals generally that…
- …has attracted significant attention in the scientific literature …
- These observations show conclusively that…
- In summary, this paper introduced the basic concepts of…
- …as can be seen from Figure # (Table #, Equation #)

If you describe someone else's results, state the results (the agent) explicitly. Don't make the research group the agent.[44]

More importantly, it has also been noted that scarless wound healing occurs in the presence of low collagen and high HA content.[44] Peramo and Marcelo (2010, p. 2021)	More importantly, wounds heal without scarring when the HA content is high and the collagen content is low.[44]

Redundant Words

To determine if a word is redundant, check if the sentence's meaning changes when you remove it. If the meaning remains the same, it's redundant. Can innovation be anything but new? (If you are talking about older innovations, then the adjective *older* is appropriate.)

Haswell, the fourth-generation Intel Core Processor, delivers a family of processors with new innovations.[1,2] Hammarlund et al. (2014, p. 6)	Haswell is the fourth-generation Intel Core Processor.[1,2]

This was the paper's first sentence. An innovation, by definition, is "a new idea, device, method" (Merriam-Webster, 2014), so qualifying one as *new* is redundant. We needn't even mention innovation: Had Haswell delivered a family of bland processors, no journal would have accepted this paper (nor would the general public have bought them).

Nowadays, you needn't specify you're talking about *nowadays*. Furthermore, a procedure or device *currently in use* might not be in use 20 years after the paper has been published.

However, most of the implants currently in use are biocompatible and have negligible toxicity. Peramo and Marcelo (2010, p. 2015)	However, most implants are biocompatible and nontoxic.

Don't single out your references. Rohrer et al. (1988) documented the data. Stating that they *already documented it in a previous study* is redundant. Including the citation demonstrates that it is documented. Saying that it is already documented, well, that is obvious. Finally, *previously* and *already* are synonyms. The published sentence has 25 words. We reduced it to 14 words.

Subsequently, the NGF antibody does indeed eliminate significant numbers of DRG neurons (about 30%), as already documented in a previous study (Rohrer et al., 1988). Gaese et al. (1994, p. 1617)	Then, the NGF antibody eliminates about 30% of DRG neurons (Rohrer et al., 1988).

Here is an example where the authors state that they note something. Is it really necessary to explicitly state that they note what follows? Can we misconstrue it as someone else noting something? Don't *note* anything. The sentence also hedges. Saying the results can be considered consistent with the model means they can also be considered inconsistent with it. Is that really what the authors wanted to say?

We note that the present interpretation of the results can be considered consistent with the so-called H-deficit model.[48] Acco et al. (1998, p. 12862)	Our results agree with the H-deficit model.[48]

Carrying four significant figures in a number indicates that you know its value precisely. Adding adverbs such as *about, approximately,* and *around* indicates that you don't know its value precisely. Converting degrees Celsius to kelvin overstates certainty when you carry too many digits.

PdO is a more active catalyst for the oxidation of methane than Pd; however, above about 1023 K, PdO begins to decompose to form Pd (Farrauto, Hobson, Kennelly, & Waterman, 1992).	PdO oxidizes methane more easily than Pd but it reduces to Pd above 750 °C (Farrauto et al., 1992).
Hayes et al. (2001, p. 4816)	

Presumably, the authors meant that PdO reduces to Pd at 750 °C, which implies an uncertainty of ±5 °C. Writing 1023 K implies an uncertainty an order of magnitude lower (±0.5 °C).

Rabinowitz and Vogel (2009) list pages of jargon and their concise scientific equivalents: *fewer in number* is *fewer*; *has the potential to* is *can*; *owing to the fact that* is *because*. They include 44 pages of misused and confused words and correct them.

State Facts

Reformulate sentences where someone states facts. You have no need to say that someone said something. Just say what they said, and cite the person. When you describe what someone did, however, put that person's name.

Lord Rayleigh (Robert John Strutt) (1) noted the paradox posed by pallasite meteorites: Olivine and metal seemingly should have separated into layers in their parent body.	The paradox of pallasite meteorites is that olivine and metal should have separated into layers in their parent body (1).
Tarduno et al. (2012, p. 939)	

Lord Rayleigh merely noted a paradox, so we can cite his paper but not that he noted something. If it were a literary work, stating his name would be appropriate.

Don't say that data shows something. State what it shows, and if you're citing a paper, include it in the references. State the evidence, not that the *evidence appears to indicate*, or *strongly suggests*.

Extensive evidence demonstrates sustained activation in frontal and parietal areas during memory delay periods.	Frontal and parietal areas remain active during memory delay periods (1-4).
Salazar et al. (2012, p. 1097)	

Empirical studies have shown that recent ocean warming has driven changes in productivity (4), population size (5), phenology (6), and community composition (7).	Ocean warming has changed productivity (4), population size (5), phenology (6), and community composition (7).
Thomas et al. (2012, p. 1085)	

Don't Say They're Facts

In all cases is a long version of *always*. *In the past* is a long version of *before*. *Despite the fact that* equals *although*. You should never use *The fact that*.

In all cases photons 1 and 4 emerge entangled despite the fact that they never interacted with one another in the past. Pan et al. (1998, p. 3892)	Photons 1 and 4 always emerge entangled although they never interacted with one another before.

2.2 PRECISE VERSUS VAGUE

Be precise (Scott, 1989; Strunk and White, 2000). Be quantitative, not qualitative and don't exaggerate (Strunk and White, 2000).

Personal pronouns shorten sentences. *It is presumed* sounds presumptuous. It is imprecise—who presumes?

It is presumed that regions with 'anomalous' conformations are present in small amounts compared with the bulk DNA (which may be in a typical DNA B configuration). Chan and Wells (1974, p. 205)	We presume that …

How small is small? It is better to say that the amounts were smaller than something or give a value (and then qualify that value with an adjective, if you wish).

To assess is better than *for assessing* which is better than *for making an assessment*. *Efficiency* is a noun, and *efficient* is an adjective. *Is efficient* is better than *to have efficiency*. When you qualify something, use an adjective. The method is sensitive; this is qualitative. Who knows how sensitive *sensitive* is? Instead of introducing the method with metadiscourse, we gave the method. In the corrected sentence, we can be vague with *little* because the authors give the data later.

A sensitive method for assessing if two amplicons have the same efficiency is to look at how ΔC_T varies with template dilution. Livak and Schmittgen (2001, p. 403)	Two amplicons are equally efficient if ΔC_T varies little with template dilution.
The neurotrophin-independent survival of NG neurons, which extends over a particularly long period of time compared with other cranial ganglia, has been correlated with the comparatively long distances that the axons have to cover to reach their targets (Vogel and Davies, 1991). Gaese et al. (1994, p. 1616)	NG neurons survive longer without neurotrophin than other cranial ganglia because the axons must travel further to reach their targets (Vogel and Davies, 1991).

Vague statements are frustrating. *An improvement* without data to substantiate your claim is pointless. It annoys readers and wastes space. *Improved speed* or *faster*? There is no need to tell us it's an improvement. If you need to specify this, reformulate your sentence: *It's two times faster.*

So, be quantitative—cite the data—and forget the qualifiers and the boosters. Qualitative statements are appropriate either when everyone understands the magnitude of the actual values or when the values are unimportant. But keep the big guns holstered: *extremely*, *incredibly*, even *very*. These adverbs are often exaggerations. You can give us percent increases, such as twice as fast, half as much, etc. Ransom's rules for technical and scientific writing state that if it is ambiguous, it is wrong (Davis, 1997). And if it's vague, it's ambiguous, therefore it's wrong.

In an article we reviewed in 2013, the authors wrote *huge* five times on the first page. How much is huge? Is the author's huge the same as everyone else's? Sometimes, huge may be just 1%.

Englander (2014) states that boosting—introducing words such as *huge, extremely, undoubtedly, obviously*—"creates emphasis and certainty." These words are extraneous. "Data talks, bullshit walks" (Patience et al., 2013). Data doesn't argue or persuade "that there is no alternative conceptualization of the world" (Bazerman, 1988). Claims authors make are not factual. Data is factual. Quantitative values are better than qualitative statements. Leave boosting for romance novels. We rephrase expressions containing the emphasized adverbs in the left-hand column:

is *extremely* dependent on temperature	depends on temperature
increases *significantly* with temperature	doubles with every 10 °C rise
samples were withdrawn *very* often	we withdrew two samples per hour
the test lasted *approximately* 5 s	the test lasted 5 s

Be specific: instead of stating that samples were withdrawn often, cite the frequency.

Be quantitative: *rates double* is better than *rates increase a lot*, or *significantly*. Adjectives are appropriate when you have defined them. A parenthetical expression may follow a quantitative statement to emphasize the authors' perception of the phenomenon.

Adverbs such as *approximately*, *about*, and *around* express uncertainty. Don't use them if you have already expressed the uncertainty or if your numerical value has three or more significant figures. In some patents, these words express an uncertainty of 5% to 10%. If a pump costs $103 000, you can say that it costs about $100 000. When you define a cost with three significant figures, the word *about* is inappropriate.

Be careful to correctly apply adjectives to physical quantities such as height, value, and time:

- Bigger, larger (physical size)
- Higher (height, position)
- Greater (quantity, value)
- Longer (time, length)

State What Is, Not What Isn't

Positive words and statements convey ideas more easily (Strunk and White, 2000). People assimilate text better when you structure it in the positive sense, so minimize the word *not*. This applies to prefixes (Table 2.1): *insufficient* is better than *not sufficient*. Adding adjectives for emphasis is often unwarranted.

Don't infirm, affirm. A negative in a contrasting sentence such as *Don't infirm, affirm* is good, though (Strunk and White, 2000). When positive words are inadequate or nonexistent, reformulate the sentence or express the idea with a negative sentence (particularly imperative sentences): How else could we say *don't use qualifiers? Use no qualifiers? Avoid qualifiers?* They aren't the same. If you can find a word that is the exact antonym of the word you are using, make your sentence positive.

TABLE 2.1 Prefixes

Weak	Better
Not accurate	Inaccurate
Not correct	Incorrect
Not complete	Incomplete
Not favorable	Unfavorable
Not sufficient	Insufficient
Not necessary	Unnecessary
Not sensitive	Insensitive

2.3 ACTIVE VERSUS PASSIVE

Four important grammatical terms are *subject*, *object*, *patient*, and *agent*. Sentences contain at least a *subject* and a verb. Most sentences also have *objects*. Normally, the *subject* goes before the verb and the *object* follows it.

A scientist (*subject, agent*) wrote (verb) this book (*object, patient*).

An *agent* performs an action. A *patient* is the target of the action. In active sentences, such as the one above, the *subject* is the one performing the action (*to write*) on the *object*. So, the *subject* is the *agent* and the *object* is the *patient*. The active voice is clearer and shorter than the passive voice.

In passive sentences, the subject still precedes the verb and the object follows it. What changes versus active sentences is that the subject is now the patient and the object is the agent.

No, the book (*subject, patient*) was written by engineers (*object, agent*).

These sentences (when the passive verb is the sentence's main verb) need glue (zero) words whose only purpose is to make the sentence grammatically correct. They convey no additional meaning. When you make passive sentences active, you remove them and increase meaningful word density. Glue words include *by* and conjugations of the verb *be*.

The passive construction is fine when you don't know the agent or the agent is unimportant (Burger Associates Inc., 1991):

The void space of the vessel was filled with corundum beads.

Presumably, the experimenter or an employee filled the vessel; it is unimportant who did it, so in this case the passive is acceptable. However, you can improve this sentence with the active voice:

Corundum beads filled the vessel void space.

Explicit Agent

When sentences contain the agent or agents, you can easily correct them. You need but swap agent and patient: make the agent the subject and the patient the object. Agents often follow the word *by*.

All the registers can be accessed not only by SIMD instructions but also by scalar instructions. 　　　Yoshida et al. (2013, p. 17)	SIMD and scalar instructions can access all registers.

The agents are *SIMD instructions* and *scalar instructions*, the passive verb is *can be accessed*, and the patient is *all the registers*. We move the agents to the beginning of the sentence and the patient to the end. Emphasizing that scalar instructions can now access all registers is unnecessary. Allow readers to make up their minds about what is important and what is unimportant. We can use the common element *instructions* to remove a word.

Better Agent in Sentence

In the experimental section, sentences often contain a better agent than the implicit *we*: the instrument.

The painted specimen was then dried in a furnace at 120 °C for 1 h. 　　　Man et al. (2002, p. 230)	A furnace then dried the painted specimen at 120 °C for 1 h.

The implicit agent is *we*: the authors dried the specimen with the furnace. But it's really the furnace that did the action, so make the furnace the agent.

An alternative with intransitive verbs is to use the intransitive form. Transitive verbs take an object, intransitive ones don't. Remove the auxiliary and you make the patient the agent.

The catalyst was calcined at 573 K for 4 h in air. Akita et al. (2005, p. 268)	The catalyst calcined at 573 K for 4 h in air.

References

When the agent is implicit but it's a reference, you can make the authors of the paper you reference the agent.

It has been suggested[48] that two types of unhydrogenated sites exist in *a*-Si. Acco et al. (1998, p. 12862)	Zahar and Shiff[48] suggested that *a*-Si has two types of unhydrogenated sites.

Saying *it has been* is weak, just like *it is* and *there are*. The agent is the reference's authors. State it explicitly. The active sentence sounds less presumptuous. In the passive version, the authors immediately cite who suggested that there were two types of sites and don't even use the active voice. Why? Habit. A better sentence states the facts directly:

a-Si has two types of unhydrogenated sites.[48]

Authors

When the implicit agent is the paper's authors, say *we*. Sometimes you can get by without it, but when you need it, say it. This contradicts your earliest introduction to technical writing that insisted the passive impersonal voice was authoritative and disengaged. It is time to change these perceptions.

Similar results were obtained when Sox-2-Cre mice were injected in the same locations (table S2). Friedmann-Morvinski et al. (2012, p. 1082)	When we injected virus into Sox-2-Cre mice (with...) in the same places (table S2) the results were similar.

The authors obtained the results when they injected mice with virus. The active sentence has just as many words as the passive sentence, but it's more direct. The authors say what they did.

Although saying *we* is better than using the passive voice, it's even better if you reformulate the sentence so you don't need to say *we*. Reformulating the sentence could mean as little as changing the verb.

The methanol conversion over 90% was achieved between 250 and 290 °C with a constant feed flow rate of 1.65 ml h^{-1}. Won et al. (2006, p. 162)	Methanol conversion exceeded 90% between 250 and 290 °C at a flow rate of 1.65 ml h^{-1}.

If we substitute *exceeded* for *was achieved* (and move the words around a little), the sentence is active without *we*.

Remove passive verbs that add nothing to the meaning of sentences.

The smallest tube observed was 2.2 nm in diameter and was the innermost tube in one of the needles (Fig. 1*c*). Iijima (1991, p. 56)	The smallest tube was 2.2 nm in diameter and was the innermost tube in one of the needles (Fig. 1*c*).

If we remove *observed*, the sentence is still grammatically correct, and it is better. *Observed* implies that someone examined the tube. So, you could write *We observed*, but it is unnecessary to change the agent from the *tube* to *we* because it is a fact that the smallest tube was 2.2 nm in diameter.

Readers

Journal publishers don't like the personal pronoun *you*. So, when you talk to readers, use the imperative mood, but the personal pronoun *we* is also acceptable. The passive voice is inappropriate, don't use it.

Each 32-nucleotide brick is a modular component; it binds to four local neighbors and can be removed or added independently. Ke et al. (2012, p. 1177)	Each 32-nucleotide brick is a modular component; it binds to four local neighbors, and we can add or remove it independently.

This sentence talks to readers. It would read better as *and you can add or remove it*, but since journal publishers don't like the word *you*, use *we*. Note that *add or remove* sounds better than *remove or add*.

When you recommend that the readers do something, *we* encompasses the readers. In fact, it sounds like *if anyone wants to do this, he or she must do so and so*.

The GIA must therefore be considered when estimating ice-sheet mass balance with either technique. Shepherd et al. (2012, p. 1184)	Therefore, we must consider the GIA when we estimate ice-sheet mass balance with either technique.

Others

Sometimes the agent includes a community of researchers or it could be common knowledge. In either case, the agent is not a distinct agent that you can reference explicitly, so the passive is acceptable. When you can reformulate the sentence to remove the passive verb without having to mention the agent, do so.

> Most have isotopic ratios that fall near the terrestrial mass fractionation line and are called *main group* pallasites (4).
> Tarduno et al. (2012, p. 939)

> Most are "main group" pallasites: their isotopic ratios fall near the terrestrial mass fractionation line (4).

We could insert the agent *researchers*, but we can reformulate the sentence to remove the passive verb. In fact, we could remove *called* and the sentence would be active and still make sense, but reformulating it makes it sound better.

If a sentence doesn't contain its agent, you must find it. When it's a reference, this is easy. Sometimes it can be difficult, so don't make your readers do it. Make the agent clear.

> Sparc64 VIIIfx was designed for the K computer.
> Yoshida et al. (2013, p. 17)

> Fujitsu designed Sparc64 VIIIfx for the K computer.

This sentence referenced nobody. However, since Fujitsu designs Sparc, Fujitsu probably designed Sparc64 VIIIfx. Inserting the agent costs no words, but the active sentence is clearer because the agent is explicit.

2.4 VIGOROUS VERSUS FEEBLE

Don't be unsure of everything you say, as if you're preparing an excuse; this is hedging. Initial results may appear to potentially indicate that perhaps there exists a possibility, remote though it may be, that maybe this sentence is bad, but we're unsure.

> Preliminary results appear to indicate that the delivery of a liquid phase containing preparations of 0.05 of dermatan sulfate and 0.2 wt% of HA may be an optimal method for use at the skin-device interface[109] (Fig. 4).
> Peramo and Marcelo (2010, p. 2022)

> An aqueous phase with a mass fraction of 0.05% dermatan sulfate and 0.02% HA regenerates the skin at the interface with the device better than other methods.

Don't say *results indicate*; it's hedging. Instead, just state your conclusions. Avoid constructions such as *for use at*. The corrected sentence is still flawed. *Better than* is vague. We don't know how much better this method is than other methods. Quantify your statements.

> Thus, such thin features, although obeying our design criteria, appeared to be less stable than were the better-supported or thicker features.
> Ke et al. (2012, p. 118)

> Thus, although such thin features obeyed our design criteria, they were less stable than better-supported or thicker features.

To say that something *appears to be* is hedging.

Convey important actions or ideas with verbs, not nouns. You mutilate (Burger Associates Inc., 1991), smother (Scott, 1989), nominalize (Pinker, 2014), zombify (Sword, 2012), or ionize the main verb when you convert it to a noun. Just as negatively charged ions are larger than their parent molecule, ionized verbs—nouns made from verbs—are also larger. Sentences containing ionized verbs are longer and vaguer. Discharge the ions and let the verb tell the story.

Here we investigated the interaction of protein kinase CKII (casein kinase II) and β-catenin. Bek and Kemler (2002, p. 4743)	Here we investigated how protein kinase CKII (casein kinase II) and β-catenin interact.

The original sentence is weak: the authors investigated an interaction, but what does that mean? Did they investigate why the kinase and catenin interact? How they interact? Even if they investigated all these questions, it's clearer to state them. If readers have to guess, you have failed to properly convey the information.

Since sentences need verbs to be grammatically correct, sentences containing an ionized main verb will also contain a helper verb. Helper verbs are only there to make the sentence grammatically correct. Linguists refer to this construction as periphrasis.

Some verbs and their ions are as follows:

consider	take into consideration
examine	undertake an examination of
present	give a presentation of
produce	production of
operate	perform an operation

Other words that people have a tendency to ionize include *activate, add, approximate, calculate, convert, degrade, describe, discuss, form, interpret,* and *observe.*

The addition of RavZ to isolated liposomes (fig. S5) drove the complete conversion of lipidated GR-PE back to an apparently unmodified form of GR (Fig. 3C). Choy et al. (2012, p. 1074)	Adding RavZ to isolated liposomes (fig. S5) converted all the lipidated GR-PE to an apparently unmodified form of GR (Fig. 3C).

A new method for the production of nano-sized materials is presented.
  Maggioni and Castagliuolo (2014,
                    p. 489)
	We present a method to produce nano-sized materials.

Give can be a helper verb.

A detailed description of the conventional lactic acid process has been given by Datta et al.[1] Mäki-Arvela et al. (2014, p. 1911)	Datta et al.[1] described in detail the conventional lactic acid process.

Do you *give a description* or do you *describe*? Which is superior? The former contains three times as many words. This is periphrasis. *Make a measurement* and *measure* is another example of periphrasis.

    Be can also be a helper verb.

First, the formation of interface states is suppressed in graphene-semiconductor junctions (*11*) because the interaction between chemically inert graphene and a completely saturated semiconductor surface—that is, one without dangling bonds—is negligible (*12*). Yang et al. (2012, p. 1140)	First, interface states do not form in graphene-semiconductor junctions (*11*) because chemically inert graphene interacts negligibly with completely saturated semiconductor surfaces—that is, without dangling bonds (*12*).

The mutilated verbs are *form* and *interact*. We made *semiconductor surfaces* plural because, since it's not a specific semiconductor surface, we save a word.

For the $\Delta\Delta C_T$ calculation to be valid, the amplification efficiencies of the target and reference must be approximately equal. Livak and Schmittgen (2001, p. 403)	To correctly calculate $\Delta\Delta C_T$, target and reference amplification efficiencies must be about equal.

Some helper (crutch) verbs are very common. *Use* is one of them. Whenever you see it, check for a better verb. Minimize (eliminate) *used* in text.

...degenerate primers were used successfully to characterize over 160 diverse aroA-like sequences ... Inskeep et al. (2007, p. 934)	...degenerate primers characterized over 160 diverse aroA-like sequences...

Two methods were generally used to improve processibility of PEEK in previous works. Li et al. (2014, p. 40595:1)	Previous work improved PEEK processibility in two ways.

Another deficiency of the sentence above is the word *works* referring to previous studies, articles, or research. The correct form is *work*.

> The satellite data sets were developed by using independent methods and, in the case of the LA, gravimeter, and SMB data sets, through contributions from numerous research groups.
> Shepherd et al. (2012, p. 1184)

> Other research groups contributed LA, gravimeter, and SMB satellite data sets. We developed the rest with independent methods.

If you say what a research group did, make the research group the subject. If you use numerical citations, state the authors; don't say the number did it (Kohler, 2014). If there are more than two or three references, use a noun such as *researchers*. Pronouns can help.

> The approach has the advantage of allowing changes in SMB and ice dynamics to be examined separately at the scale of individual glacier drainage basins (5) and has been used in numerous assessments of AIS and GrIS mass balance (*18, 55-57*).
> Shepherd et al. (2012, p. 1185)

> With this approach, we examine changes in SMB and ice dynamics separately at the scale of individual glacier drainage basins (5). Researchers have assessed many AIS and GrIS mass balances with it (*18, 55-57*).

The helper here is *occurs*.

> No conformational change occurs to AtCERK1-ECD after chitin binding.
> Liu et al. (2012, p. 1161)

> AtCERK1-ECD's conformation does not change after binding to chitin.

Writing with crutches such as *occurs* is weak. People use them when they can't find a verb that describes the action. *Change* is a verb; *a change* is a noun. The action here is actually *change*. If we just use the noun *conformation*, the sentence is better. We also added *to*. This may seem like an innocuous change, but it clarifies the sentence. *Chitin binding* is vague. Readers inexperienced in the field don't necessarily know if AtCERK1-ECD binds to chitin or if chitin binds to something else. Although if you read the rest of the paper you will understand that it's AtCERK1-ECD that binds to chitin, when you read the sentence alone you can't know this for sure. In this example, *binding* is a noun. When we make it a verb, we have to add *to*, but the sentence becomes much clearer.

The helper here is *have*.

> Temperature errors have two types of *effects* on kinetic parameters and the resulting kinetic predictions [11].
> Vyazovkin et al. (2011, p. 5)

> Temperature errors *affect* kinetic parameters and the resulting kinetic predictions in two ways [11].

Effects mutilates *affect*.

The structure *noun-ending-in-ion of* is weak. Just make the noun a verb ending in *ing*.

The tight packing of the three LysMs suggests that deletion of one LysM could generate a deleterious effect on the structural integrity of the other two. Liu et al. (2012, pp. 1160-1161)	The three LysMs are tightly packed, so deleting one could damage the remaining two's structural integrity.

Deletion of is *deleting*. Never write *deletion of*. *Generate a deleterious effect on* is a verbose alternative to *damage*. We referred to a thesaurus and replaced five words with one.

Direct calculation of spectra from a six-dimensional surface is unnecessarily difficult. Jankowski et al. (2012, p. 1147)	Directly calculating spectra from a six-dimensional surface is difficult.

Did a specific individual make it difficult to calculate the spectra? How can they say it's unnecessarily difficult? This is self-conscious. Don't complain about how difficult your job is.

This sentence has many deficiencies: loss, deletion, leaves.

Thus, loss of this base pair or deletion of G1 leaves aminoacylation virtually unaffected. Jahn et al. (1991, p. 258)	Thus, losing this base pair or deleting G1 does not affect aminoacylation.

The following sentence highlights a problem with verb mutilation: even though there is no agent, we don't need a passive sentence.

Dehydroxylation performed at high temperature leads to creation of Lewis acid sites. Mäki-Arvela et al. (2014, p. 1919)	High temperature (800 °C) dehydroxylates La(OH)$_n$ creating Lewis acid sites.

When you read this sentence out of context, you have no idea what is dehydroxylated. It is missing a pronoun to refer to the previous sentence that has the agent. This sentence is imprecise. We improved it by eliminating the crutch verb perform and specifying the temperature at which the agent - La(OH)n - is dehydroxylated.

2.5 STRAIGHTFORWARD VERSUS CONVOLUTED

When you write simply, you have a better chance of writing correctly. Complicating a sentence is easy. Just fish in the thesaurus for synonyms, mutilate a few verbs while you're at it, qualify your nouns and verbs a little, and before long you've got a sentence longer than this one that conveys too many ideas and that no one but a compiler can parse. On the other hand, you can't remove words forever without changing the meaning of a sentence. Short sentences are more likely to be good than long ones.

The average sentence length is 15 words (Scott, 1989), but some sentences have as many as 60 words. The out-of-breath rule tests for long sentences: Read the sentence out loud without stopping to take a breath. If you're out of breath at the end, the sentence is too long. Within a few seconds, if you don't understand the sentence, it is too long. When you forget how the sentence started, it is too long.

As a general rule, keep every sentence to one idea (Scott, 1989). Covering more ideas per sentence dilutes the message. The more complicated the subject, the shorter sentences should be (Scott, 1989).

Long sentences become ambiguous more easily than short sentences. And if they're ambiguous, they're wrong (Ransom's rules; Davis, 1997). If your reader has to guess, you've made a mistake.

> *Here, we compare and combine estimates of ice-sheet mass balance derived from all three satellite geodetic techniques, using common spatial and temporal domains, to investigate the extent to which the approaches concur and to produce a reconciled estimate of ice-sheet mass balance.*
>
> Shepherd et al. (2012, p. 1184)

This sentence has 42 words: it approaches the 2σ boundary for *Nature's* top 500 articles of 2013 (25 words per sentence and $\sigma = 10$). In 2013, only 1% of the sentences in *Science* abstracts had more than 43 words.

Split sentences up when *which*, *that*, and *and* make them long.

The autophagy pathway is used by eukaryotic cells to sequester cytosolic proteins and organelles into a membrane-bound compartment called an autophagosome (AP), which fuses with lysosomes to promote cargo degradation (5).	Eukaryotic cells sequester cytosolic proteins and organelles into an autophagosome (AP) through the autophagy pathway. The AP, a membrane-bound compartment, fuses with lysosomes to degrade cargo (5).
Choy et al. (2012, p. 1072)	
Atg4 is a cysteine protease that regulates autophagy by cleaving the C terminus of pro-Atg8 after the conserved glycine, which is required for the glycine to participate in the PE conjugation reaction (5).	The cysteine protease Atg4 cleaves the C terminus of pro-Atg8 after the conserved glycine to regulate autophagy. The glycine can then conjugate with PE (5).
Choy et al. (2012, p. 1074)	

| The slightly less periodic (higher COV) San Andreas fault is known to be influenced by major faults nearby (*30*), and the aperiodic Dead Sea Transform is a more complicated system and slower slipping (*9*) than either the San Andreas or the Alpine Fault.
Berryman et al. (2012, p. 1693) | Nearby major faults influence the less periodic (higher COV) San Andreas fault (*30*). The aperiodic Dead Sea Transform is more complicated and slips more slowly (*9*) than either the San Andreas fault or the Alpine Fault. |

To improve the long sentence above, we remove the conjunction and make two sentences. Next, we replace the passive verb *is known to be influenced* with *influence*. To do this we have to invert the agent/patient order so that *nearby major faults* becomes the agent and the San Andreas fault becomes the patient. We remove *is known to be* because a reference follows the information. *Slightly* means nothing unless the authors define it somewhere in their paper. Use verbs when they're better: *slips* more slowly is better than *is slower slipping*.

Bullet points for lists of two or more items when the items are long are effective (Burger Associates Inc., 1991).

● The channel consists of two modules: (i) a stem that penetrates and spans a lipid membrane, and (ii) a barrel-shaped cap that adheres to the cis side of the membrane.

Langecker et al. (2012, pp. 932-933)

Split sentences into parts and introduce words such as *however* and *therefore* occasionally.

| Although a consequence of this revision is a potential discrepancy between far-field sea-level records and commonly accepted Northern Hemisphere deglaciation models, both of the new regional GIA models perform well when compared with Antarctic Global Positioning System (GPS) observations (*34*), and we conclude that these latest solutions are best suited for estimating AIS mass balance.
Shepherd et al. (2012, p. 1184) | This revision may cause a discrepancy between far-field sea-level records and standard Northern Hemisphere deglaciation models. However, both new regional GIA models compare well with Antarctic Global Positioning System (GPS) observations (*34*). Therefore, these latest solutions estimate AIS mass balance best. |

2.6 EXERCISES

Simplify the following sentences and phrases.

1. It is interesting to remark that these results indicate that deactivation phenomena are not important below 500 °C for PdO/ZrO$_2$-Y. (Escandón et al., 2005)

2. The adsorbate is represented by the filled circles and the Ni by open circles. (Dudzik et al., 1998, p. 12659)

3. The CPUs are connected by directory-based cache coherency links to minimize the frequency of the snoop transactions. (Yoshida et al., 2013, p. 20)

4. In Desharnais et al. [2000, 2002], a theory of approximation for LMPs was initiated and was refined and extended in Danos and Desharnais [2003] and Danos et al. [2003]. (Chaput et al., 2014, p. 5:2)

5. The procedure is repeated until we reach the sphere E. (Gotoh and Finney, 1974, p. 203)

6. So, if the present structure is considered, at least four main zones can be distinguished (from east to west: Hellenic arcs; Carpathian arcs; Alpine arcs; and Apennine-Tell-Rif arcs). (Boccaletti and Guazzone, 1974, p. 18)

7. The Apuseni Mountains can also be considered as a remnant arc containing the more internal parts of the inner Dinarids. (Boccaletti and Guazzone, 1974, p. 19)

8. Although the specific role of these activity patterns is not fully understood, theoretical, anatomical, and electrophysiological studies suggest that synchronous interactions among these cortical regions support working memory processes (*5–11*). (Salazar et al., 2012, p. 1097)

9. Isoconversion methods can be categorised into one of two main groups of methods. (Starink, 2003, p. 164)

10. Others[3,4] have found that MDCK cells transfected with the genes for foreign secretory proteins release these polypeptides both apically and basolaterally. (Caplan et al., 1987, p. 632)

11. Sparc64 VI and Sparc64 VII were used in a Unix server called Sparc Enterprise.[7,8] (Yoshida et al., 2013, p. 17)

12. Then, the voxel information of multiple shapes is interpreted by a custom program to generate a list of strands involved in the formation of each shape. (Ke et al., 2012, p. 1181)

13. These defence systems rely on small RNAs for sequence-specific detection and silencing of foreign nucleic acids. (Jinek et al., 2012, p. 816)

14. The basic procedure at this stage is to use a series of pairwise alignments to align larger and larger groups of sequences, following the branching order in the guide tree. (Thompson et al., 1994, p. 4675)

15. Clays are also suitable support materials exhibiting acidity, and their properties can be tuned by addition of metal. (Mäki-Arvela et al., 2014, p. 1919)

16. One set of methods relies on approximating the so-called temperature integral and requires data on $T_f(\beta)$ only. (Starink, 2003, p. 164)

17. Radiocarbon dating of collagen[3] from the adult mammoth was undertaken both by accelerator mass spectrometry at Oxford University, and by proportional counting of CO_2 at Birmingham University. (Coope and Lister, 1987, p. 474)

18. Use of optical traps for the manipulation of biological particles was recently proposed, and initial observations of laser trapping of bacteria and viruses with visible argon-laser were reported.[1] (Ashkin et al., 1987, p. 769)

19. It is now known that icosahedral symmetry is compatible with quasiperiodic translational order, and states of matter arranged in such a way have been found (i.e., quasicrystals[2]). (Cozzini and Ronchetti, 1996, p. 12040)

20. For example, warming-induced thinning of Jakobshavn Isbræ's floating ice tongue eventually led to its complete loss and faster terminus flow with no sustained regrowth of the tongue (*42*), despite the fact that the potential melting rates even with Irminger Water in the fjord would allow persistence of a reduced ice shelf capable of at least some buttressing. (Joughin et al., 2012, p. 1175)

21. Helium is used as a critical tracer throughout the Earth sciences, where its relatively simple isotopic systematics is used to trace degassing from the mantle, to date groundwater and to time the rise of continents[1]. (Lowenstern et al., 2014)

22. At room temperature, the resistance of both modes is found to increase abruptly at a particular length—the ballistic mode at 16 micrometres and the other at 160 nanometres. (Baringhaus et al., 2014).

23. Here we report the achievement of fusion fuel gains exceeding unity. (Hurricane et al., 2014)

24. The observed spatial distribution rules out symmetric explosions even with a high level of convective mixing. (Grefenstette et al., 2014)

25. operating at a pressure of 3 atm and a temperature of 390 °C

26. Diffusion limitations around the crucible can be neglected

27. We analyzed the effect of flow rate and found that when the flow was 20 nL/h or higher, there was no effect on the reaction rate.

28. More studies should be done in order to have a better understanding

29. Xylose dehydrates to furfural in the presence of different types of acid catalyst

30. We studied the effect of temperature and percentage of oxygen on reaction rate.

31. Due to the extremely high heat transfer rates

32. The data in the table show a good correlation

33. Heated in the presence of air

34. Once the temperature achieved steady state conditions

35. plays a very important role in the phenomenon
36. the model is used to describe the different conversion profiles
37. increasing temperature gives much better results
38. and the results showed that
39. are used to represent
40. The device allowed to obtain a stable emulsion.
41. The use of the new apparatus allowed to reach very satisfactory monoclonal antibody production rates.
42. offer an alternative explanation

Chapter 3

Reporting Data

Chapter Outline

In this chapter, we review statistics and how to report experimental data precisely. You conduct experiments, repeat measurements several times, take an average, report a confidence interval et voilà. What could go wrong? Patience et al. (2013) chose one of the top three chemical engineering journals and selected 20 papers in a row in an issue from 2012. Eighteen of the 20 articles reported experimental data imprecisely or inefficiently. The two commonest errors were carrying too many significant figures, and not reporting uncertainty ($x \pm \Delta_x$). When they did report uncertainty, they neglected to state whether it represented standard deviation, standard error, or confidence intervals. Another common deficiency was expressing numbers with inappropriate unit prefixes.

$$
\begin{array}{lll}
1.23 \times 10^5 \, \text{m} & \text{vs.} & 123 \, \text{km} \\
1.6 \times 10^{-19} \, \text{J} & \text{vs.} & 1.0 \, \text{eV} \\
45\,800 \, \text{J mol}^{-1} & \text{vs.} & 45.8 \, \text{kJ mol}^{-1}
\end{array}
$$

Inadmissible, systematic, and random errors affect all data (Patience, 2013). Operating equipment incorrectly and mistyping data into tables are examples of inadmissible errors—blunders. Systematic errors relate to physical laws, and are corrected by calibrating instruments, adjusting the zero, or controlling the environmental conditions (relative humidity and temperature in an environmental testing chamber or glove box). Random errors relate to the inability to reproduce experimental data precisely for known or unknown reasons. Random errors are represented by statistics.

3.1 THE INTERNATIONAL SYSTEM OF UNITS

Most journals require that authors conform to the Système International d'Unités (SI) and they publisher's may even change units to SI during copyediting.

Communicate Science Papers, Presentations, and Posters Effectively. http://dx.doi.org/10.1016/B978-0-12-801500-1.00003-6

The SI has its roots in the late eighteenth century prior to the French revolution. It was inspired by the metric system (BIPM, 2006), and originally had two standard quantities—length and mass. The Metre Convention of 1875 formally established an institute to coordinate international metrology (BIPM, 1875). The convention established three organizations: the General Conference on Weights and Measures (Conférence Générale des Poids et Mesures—CGPM), the International Committee for Weights and Measures (Comité International des Poids et Mesures—CIPM), and the International Bureau of Weights and Measures (Bureau International des Poids et Mesures—BIPM). The CGPM includes governmental delegates of all the member states and endorses the rules on writing and presenting standardized data worldwide (BIPM, 2014). The BIPM is responsible for the SI, which unifies physical measurements across the world that give "accurate and comparable measurement results" (BIPM, 2013).

Seven mutually independent quantities comprise the SI base units (Table 3.1). All other units derive from these (Table 3.2).

TABLE 3.1 SI Base Units

Base Quantity	Base Unit	Symbol
Length	meter	m
Mass	kilogram	kg
Time	second	s
Electric current	ampere	A
Thermodynamic temperature	kelvin	K
Amount of substance	mole	mol
Luminous intensity	candela	cd

TABLE 3.2 SI Derived Units

Derived Quantity	Derived Unit	Symbol
Area	square meter	m^2
Volume	cubic meter	m^3
Speed, velocity	meter per second	$m\,s^{-1}$
Acceleration	meter per second squared	$m\,s^{-2}$
Wave number	reciprocal meter	m^{-1}
Mass density	kilogram per cubic meter	$kg\,m^{-3}$
Specific volume	cubic meter per kilogram	$m^3\,kg^{-1}$
Current density	ampere per square meter	$A\,m^{-2}$
Magnetic field strength	ampere per meter	$A\,m^{-1}$
Amount-of-substance concentration	mole per cubic meter	$mol\,m^{-3}$
Luminance	candela per square meter	$cd\,m^{-2}$
Mass fraction	kilogram per kilogram	$kg\,kg^{-1}$

TABLE 3.3 Acceptable Non-SI Units (BIPM, 2006)

Unit	Symbol	SI Equivalent
Degree Celsius	°C	$1\,°C = 1\,K + 273.15$
Minute	min	$1\,min = 60\,s$
Hour	h	$1\,h = 3600\,s$
Day	d	$1\,d = 86\,400\,s$
Degree	°	$1° = (\pi/180)\,rad$
Minute	′	$1′ = (\pi/10\,800)\,rad$
Second	″	$1″ = (\pi/648\,000)\,rad$
Ångström	Å	$1\,Å = 0.1\,nm$
Hectare	ha	$1\,ha = 10\,000\,m^2$
Liter	l or L	$1\,L = 0.001\,m^3$
Tonne, metric ton	t	$1\,t = 1000\,kg$
Bar	bar	$1\,bar = 100\,kPa$
Millimeter of mercury	mmHg	$750\,mmHg = 1\,bar$

Hertz (s^{-1}) is an example of a derived unit that is expressed in terms of one of the seven base quantities. Thompson and Taylor (2008) and Heldoorn (2007) report many derived units. The SI accepts non-SI units for temperature (Preston-Thomas, 1990), time, plane angle, area, and mass (Table 3.3). The CIPM adopted the symbol L to represent liter in 1980 to avoid confusing the lowercase letter *l* with the numeral 1 (one). In many English-speaking countries, the symbol "t" is used for the metric ton (BIPM, 2006). Thompson and Taylor (2008) require that all documents for the NIST conform to the SI. Alternative units may be included in parentheses after the SI unit if the author considers it necessary for the audience.

Here are some SI writing conventions:

- Plane angles (°, ′, ″) don't take spaces after numerals. Otherwise, a space separates numerals from unit symbols.
- Always add a space before °C but never between ° and C.
- Never report K with a ° symbol.
- Add a space before the % sign.
- Report the units with both the numerical value and the confidence interval:
  $2650\,kg\,m^{-3} \pm 20\,kg\,m^{-3}$, or
  $(2650 \pm 20)\,kg\,m^{-3}$, or
  $2650(1 \pm 0.008)\,kg\,m^{-3}$, or
  $2650(1 \pm 0.8\%)\,kg\,m^{-3}$.
- Italics are slanted characters; roman characters are upright. Put symbols representing quantities or variables in italics. Apply roman typeface to unit symbols (m, s, g), prefixes (MPa), chemical elements (He), mathematical operators, and universal constants (π, e).

- Numbers that have more than four digits before or after the decimal point should be separated by spaces into groups of three digits: 123 456.789 012. Numbers with four digits before and after the decimal don't need a space.
- Mass, mole, and volume fractions should be stated explicitly: Write *with a mass fraction of* 15%, rather than 15 wt%.
- The *Oxford English Dictionary* adopts *metre* as the spelling for the unit for length and *litre* as the spelling for the unit for volume, whereas NIST writes these as *meter* and *liter* (Thompson and Taylor, 2008). The Oxford English Dictionary favours the British English spelling but it gives the American spelling as an alternative form.
- Thompson and Taylor (2008) state that the correct symbols for parts per million (ppm) and parts per billion (ppb) are $\mu L\,L^{-1}$ or $ng\,kg^{-1}$. BIPM (2006) allows quantities to be expressed as parts per million but not as parts per billion because *billion* could represent 10^9 or 10^{12}, depending on the country.
- State explicitly if the error bars refer to sample standard deviation (s or SD), standard error of the mean (s_x or SEM), or confidence intervals (Δ_x or CI).
- atm is unacceptable as a unit to represent pressure.
- The CGPM has yet to accept week, month, or year as a base unit. Nevertheless, we suggest that the symbol y is appropriate to represent *year* when you refer to a parameter in an equation. The word year, month, or week is fine when you write out the number (five years), or refer to a date (the year 1983), or state a time span (40-year career or spanning twenty years).
- The abbreviations sec, secs, mins, hr, and hrs are incorrect to represent seconds (s), minutes (min), and hours (h).

3.2 REFRESHER ON STATISTICS

After you have eliminated blunders and systematic errors, repeat experiments to assess the variability in the data. The minimum number of repeats (sample size) depends on the research domain, the experimental complexity, and instrument precision. In *Science*, figure captions often state that the standard error of the mean was based on a minimum of three experiments (although an average of at least five is better). To have an accurate assessment of the standard deviation requires 10-20 measurements. Sample sizes for epidemiological studies can exceed several hundred thousand. Krzywinski and Altman (2013b) treat power and sample size in more detail.

The mean of an entire population (N) and of a sample of the population (n) are μ and $\bar{x}$, respectively. For n samples,

$$\bar{x} = \frac{1}{n}\sum_{i=1}^{n} x_i,$$

where x_i is the ith measurement. The (biased) variance of the entire population and the (unbiased) variance of a sample of the population are

$$\sigma^2 = \frac{1}{N}\sum_{i=1}^{N}(x_i - \mu)^2 = \frac{1}{N}\sum_{i=1}^{N}x_i^2 - \mu^2$$

and

$$s^2 = \frac{1}{n-1}\sum_{i=1}^{n}(x_i - \bar{x})^2 = \frac{1}{n-1}\sum_{i=1}^{n}x_i^2 - \frac{n}{n-1}\bar{x}^2,$$

respectively.

The probability density function $f(x)$ of a random variable X is the probability that X will take a value in the vicinity of x:

$$f(x) = \Pr(x - \epsilon \leq X \leq x + \epsilon),$$

where $\Pr(A) \in [0, 1]$ is the probability that event A will occur. Some properties of probability density functions are as follows:

- The probability that X takes a value in $[a, b]$ is the area under the curve of $f(x)$ from a to b:

$$\Pr(a \leq X \leq b) = \int_a^b f(t)\, dt.$$

- X must take a value in $[-\infty, \infty]$, so the area under all of $f(x)$ is 1:

$$\Pr(-\infty \leq X \leq \infty) = \int_{-\infty}^{\infty} f(t)\, dt = 1.$$

The cumulative distribution function $F(x)$ of a random variable X is

$$F(x) = \Pr(X \leq x) = \int_{-\infty}^{x} f(t)\, dt.$$

The probability density function of a random variable X that follows the Gaussian (or normal) distribution $\mathcal{N}(\mu, \sigma^2)$ is

$$f(x) = \frac{1}{\sqrt{2\pi\sigma^2}}e^{-(x-\mu)^2/(2\sigma^2)},$$

where μ is the mean and σ^2 the variance. X follows $\mathcal{N}(\mu, \sigma^2)$: $X \sim \mathcal{N}(\mu, \sigma^2)$.

The probability density function and cumulative distribution function of the standard normal distribution $\mathcal{N}(0, 1)$ are

$$\phi(x) = \frac{1}{\sqrt{2\pi}}e^{-x^2/2}$$

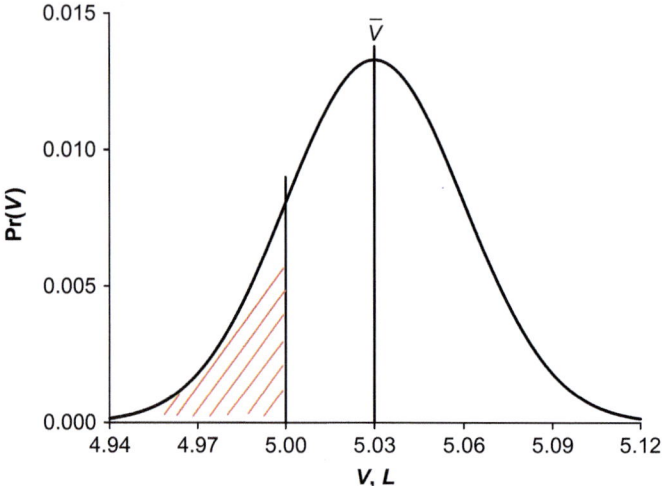

FIGURE 3.1 The probability distribution function of the volume of gasoline that a television journalist collected. The shaded area represents the probability that a gasoline pump delivers less than 5 L.

and

$$\Phi(x) = \Pr(\mathcal{N}(0, 1) \le x) = \int_{-\infty}^{x} \phi(t)\, dt.$$

$\phi(x)$ is an even function because it's symmetric around the y-axis, so $\phi(x) = \phi(-x)$ (Figure 3.1). Since $\phi(x)$ is even, $\Phi(x) + \Phi(-x) = 1$. Therefore, $\Phi(0) = 1/2$.

The cumulative distribution function of any normal distribution is directly related to the cumulative distribution function of the standard normal distribution:

$$\Pr(\mathcal{N}(\mu, \sigma^2) \le x) = \Pr(\mathcal{N}(0, 1) \le (x - \mu)/\sigma) = \Phi((x - \mu)/\sigma).$$

Rather than calculate probabilities, you can look up their values in tables. The tables relate the probability to a normalized variable z:

$$z = \frac{x - \mu}{\sigma}.$$

Since the probability distribution is symmetric around $z = 0$ and the area under the curve equals 1, the integral from $-\infty$ to 0 equals 1/2:

$$\Pr(-\infty \le \mathcal{N}(0, 1) \le 0) = \int_{-\infty}^{0} \phi(t)\, dt = \frac{1}{2}.$$

Table 3.4 gives a one-tailed distribution because it reports only half of the probability distribution function. The total probability between $-\infty$ and some

TABLE 3.4 Area Under the Curve of φ(z) from 0 to z

z	0.00	0.01	0.02	0.03	0.04	0.05	0.06	0.07	0.08	0.09
0.0	0.000	0.004	0.008	0.012	0.016	0.020	0.024	0.028	0.032	0.036
0.1	0.040	0.044	0.048	0.052	0.056	0.060	0.064	0.068	0.071	0.075
0.2	0.079	0.083	0.087	0.091	0.095	0.099	0.103	0.106	0.110	0.114
0.3	0.118	0.122	0.126	0.129	0.133	0.137	0.141	0.144	0.148	0.152
0.4	0.155	0.159	0.163	0.166	0.170	0.174	0.177	0.181	0.184	0.188
0.5	0.192	0.195	0.199	0.202	0.205	0.209	0.212	0.216	0.219	0.222
0.6	0.226	0.229	0.232	0.236	0.239	0.242	0.245	0.249	0.252	0.255
0.7	0.258	0.261	0.264	0.267	0.270	0.273	0.276	0.279	0.282	0.285
0.8	0.288	0.291	0.294	0.297	0.300	0.302	0.305	0.308	0.311	0.313
0.9	0.316	0.319	0.321	0.324	0.326	0.329	0.332	0.334	0.337	0.339
1.0	0.341	0.344	0.346	0.349	0.351	0.353	0.355	0.358	0.360	0.362
1.1	0.364	0.367	0.369	0.371	0.373	0.375	0.377	0.379	0.381	0.383
1.2	0.385	0.387	0.389	0.391	0.393	0.394	0.396	0.398	0.400	0.402
1.3	0.403	0.405	0.407	0.408	0.410	0.412	0.413	0.415	0.416	0.418
1.4	0.419	0.421	0.422	0.424	0.425	0.427	0.428	0.429	0.431	0.432
1.5	0.433	0.435	0.436	0.437	0.438	0.439	0.441	0.442	0.443	0.444
1.6	0.445	0.446	0.447	0.448	0.450	0.451	0.452	0.453	0.454	0.455
1.7	0.455	0.456	0.457	0.458	0.459	0.460	0.461	0.462	0.463	0.463
1.8	0.464	0.465	0.466	0.466	0.467	0.468	0.469	0.469	0.470	0.471
1.9	0.471	0.472	0.473	0.473	0.474	0.474	0.475	0.476	0.476	0.477
2.0	0.477	0.478	0.478	0.479	0.479	0.480	0.480	0.481	0.481	0.482
2.1	0.482	0.483	0.483	0.483	0.484	0.484	0.485	0.485	0.485	0.486
2.2	0.486	0.486	0.487	0.487	0.487	0.488	0.488	0.488	0.489	0.489
2.3	0.489	0.490	0.490	0.490	0.490	0.491	0.491	0.491	0.491	0.492
2.4	0.492	0.492	0.492	0.492	0.493	0.493	0.493	0.493	0.493	0.494
2.5	0.494	0.494	0.494	0.494	0.494	0.495	0.495	0.495	0.495	0.495
2.6	0.495	0.496	0.496	0.496	0.496	0.496	0.496	0.496	0.496	0.496
2.7	0.497	0.497	0.497	0.497	0.497	0.497	0.497	0.497	0.497	0.497
2.8	0.497	0.498	0.498	0.498	0.498	0.498	0.498	0.498	0.498	0.498
2.9	0.498	0.498	0.498	0.498	0.498	0.498	0.499	0.499	0.499	0.499

positive value, z, equals the sum of the integral from $-\infty$ to 0 (which equals 1/2) and the integral from 0 to z:

$$\Phi(z) = \int_{-\infty}^{z} \phi(t)\, dt = \frac{1}{2} + \int_{0}^{z} \phi(t)\, dt.$$

The last term is tabulated in Table 3.4.

Example. A journalist at a local television station postulated that gas station pumps were poorly calibrated in detriment to the consumer: the consumer was paying for undelivered gasoline. To test this hypothesis, the journalist collected 5 L of gasoline from 20 gas stations and measured the volume in graduated cylinders with a resolution of 1 mL (Table 3.5). (a) Calculate the mean and standard deviation. (b) What percentage of the pumps in the entire population deliver less than 5 L of gasoline if the sample is normally distributed and represents the entire population of pumps?

Solution.

(a) The sample mean is 5027.8 mL, with a sample standard deviation of 25.8 mL. Since the sample standard deviation exceeds the resolution of the graduated cylinder (1 mL), carrying five—or even four—significant figures is unreasonable. $\bar{x} = 5030$ mL and $s_x = 30$ mL. For s_x, 25 mL may be acceptable, but we truncate the values only after calculating the statistics.

(b) We want to find

$$\Pr(X \le 5000) = \Phi\left(\frac{5000 - 5027.8}{25.8}\right) = \Phi(-1.08)$$

TABLE 3.5 Volumetric Measurements of 5 L of Gasoline from 20 Gas Stations

Test	V (mL)	Test	V (mL)
1	5058	11	5010
2	5062	12	5013
3	5037	13	5028
4	5057	14	5034
5	4974	15	5020
6	5006	16	5037
7	5000	17	5046
8	5000	18	5048
9	5000	19	5057
10	5010	20	5058

but only have $\Phi(z)$ for $z > 0$. With the property we mentioned earlier,

$$\Phi(-1.08) = \Phi(1.08),$$

and from Table 3.4, $\Phi(1.08) = 0.36$. So, the area under the curve between $-\infty$ and $z = -1.08$ equals $1/2 + 0.36 = 0.86$, and the probability equals $1 - 0.86 = 0.14$. We expect 14% of the pumps to deliver less than 5 L of gasoline.

3.3   CHAUVENET'S CRITERION

When you know data is wrong because someone made a mistake during the experiment, because instruments malfunctioned, or for some other reason, eliminate it directly. Sometimes you do not know why one or more data points are noticeably different from the mean. Chauvenet's criterion states that you can discard a data point if

$$\Pr(X \leq x) < \frac{1}{2n}.$$

Example. In the previous example, 19 of the 20 samples were at least 5 L. One was 4.974 L, which is more than 2σ from the mean. Can we discard this data point?

Solution.

$$\Pr(X \leq 4974) = \Phi \left(\frac{4974 - 5028}{25.8} \right) = \Phi(-2.09),$$

$$1 - \Phi(2.09) = 0.018 < \frac{1}{2(20)} = 0.025.$$

This value falls outside Chauvenet's criterion, and in normal circumstances, we could reject it. The mean and standard deviation would become 5031 mL (5030 mL) and 23 mL (or 20 mL), respectively. We would expect only 6% not 14% of the pumps to be out of spec.

   Although we could ignore the data point thanks to Chauvenet's criterion, rejecting it defeats the purpose of the experiment, which was to detect fraud. However, calling it a fraud is premature. To exclude data from properly executed tests requires *probable cause*. We could have made an error measuring the volume of gasoline. The experiment is missing a control. We have to measure the volume of gasoline the same pump delivers many times and then calculate its standard deviation.

3.4 DEFINITIONS

Resolution	The smallest increment of a physical property that an instrument measures or displays. The resolution of graduated cylinders, beakers, and rulers is at a minimum half the distance between two hash marks.
Precision	Represents the repeatability when you measure the same physical property many times. Electronic chronometers record time with a resolution of 0.01 s, but human reaction times are on the order of 0.1 s. So, the precision of measuring with a stopwatch is 0.1 s even though the resolution of the stopwatch is an order of magnitude lower.
Accuracy	Represents how well an instrument reports the true value—a value a standard or theory certifies. Physical laws govern accuracy. To improve accuracy, we calibrate instruments, correct the zero, etc. Statistical laws govern precision.
Error	The difference between true and measured values. Systematic errors are for accuracy and random errors are for precision.
Uncertainty	A measure of your confidence that a mean value lies between two intervals that you define. It "characterizes the dispersion of the values that could reasonably be attributed to the measurand" (JCGM, 2008).

Type A uncertainty relates to statistics—a multiple of the standard deviation or a confidence interval, for example—-while type B uncertainty relates to experience, instrument resolution (e.g., graduations on a flask), the manufacturer's specification, and uncertainties from reference books (JCGM, 2008). The method of extremes is a type B measure of uncertainty:

$$\bar{x} = \frac{x_{max} + x_{min}}{2},$$

$$\Delta_x = \frac{x_{max} - x_{min}}{2}.$$

Error bars in graphs must be stated explicitly. In *Science* and *Nature*, authors usually define them as the standard deviation, σ, or the standard error of the mean, $\frac{\sigma}{\sqrt{n}}$. (JCGM (2008) defines the latter as "experimental standard deviation of the mean" and not the "standard error of the mean.")

Type A uncertainty assumes the quantity follows a uniform probability law in $[a, b]$. It defines a range of values within which the true value is likely to be. To increase your confidence that the measurand lies within an interval, multiply your statistical parameter, σ or $\frac{\sigma}{\sqrt{n}}$, by a coverage factor, $z(\alpha)$:

$$\Delta_x = z(\alpha)\sigma_x = z(\alpha)\frac{\sigma}{\sqrt{n}},$$

where $z(\alpha)$ is the z-statistic and α is the confidence level, which is 95% most of the time. In more stringent tests, α is 99% or even 99.9%.

For a normally distributed population of data, $z(95) = 1.96 \approx 2$: a single measured value has a 95% chance of lying within 2σ of the mean. The value of 1.96 comes from Table 3.4. The area under the curve (one-tailed) from 0 to 1.96 equals 0.475. So the two-tailed area under the curve, from -1.96 to 1.96, is $2 \times 0.475 = 0.95$. A value has a 68% chance of lying within σ of the mean and a 99.7% chance of lying within 3σ.

When you repeat the experiment fewer than 30 times, calculate the confidence interval based on the Student's t-statistic:

$$\Delta_x = t(\alpha, n - 1)s_x = t(\alpha, n - 1)\frac{s}{\sqrt{n}},$$

where s_x is the sample standard deviation, n is the number of repeats (sample size), and $n - 1$ is the number of degrees of freedom. Rather than reading tables for the t-statistic, we approximate $t(0.95, n)$, $t(0.99, n)$, and $t(0.999, n)$ by

$$t(0.95, n) = -0.441 + 2.40n/(n - 1.01) \quad \forall n \geq 3,$$

$$t(0.99, n) = -0.653 + 3.23n/(n - 1.56) \quad \forall n \geq 5,$$

$$t(0.999, n) = -1.08 + 4.36n/(n - 2.29) \quad \forall n \geq 7.$$

Example. If the 19 gasoline samples from the previous example were from the same pump, what is the 95% confidence interval (uncertainty)?
Solution.

$$\Delta_V = t(0.95, n - 1)\frac{s}{\sqrt{n}},$$

$$t(0.05, 18) = -0.441 + \frac{2.40n}{n - 1.01} = -0.441 + \frac{2.40 \times 18}{18 - 1.01} = 2.102.$$

$$\Delta_V = 2.101\frac{23}{\sqrt{19}} = 11.$$

The value of the t-statistic from Table 3.6 is 2.101, which compares well with 2.102 derived from our equation.

One way to express the volume of gasoline is $5030 \, \text{mL} \pm 10 \, \text{mL}$. The confidence interval Δ_V is half the standard deviation: we are 95% confident the true value of the mean lies between 5020 and 5040 mL.

JCGM (2008) prefers one of the following to state uncertainty:

1. $V = 5.03 \, \text{L}$ with an uncertainty $\Delta_V = 0.01 \, \text{L}$.
2. $V = 5.03(1) \, \text{L}$, where the number in parentheses is the numerical value of Δ_V of the last digit of the quoted result.
3. $V = 5.03(0.01) \, \text{L}$, where the number in parentheses is the numerical value of Δ_V expressed in the unit of the quoted result.
4. $V = (5.03 \pm 0.01) \, \text{L}$.

TABLE 3.6 Student's *t*-Statistic

n	t(0.05, n)	t(0.01, n)	t(0.001, n)
1	12.71	63.66	636.6
2	4.303	9.925	31.60
3	3.183	5.841	12.92
4	2.777	4.604	8.610
5	2.571	4.032	6.869
6	2.447	3.707	5.959
7	2.365	3.500	5.408
8	2.306	3.355	5.041
9	2.262	3.250	4.781
10	2.228	3.169	4.587
11	2.201	3.106	4.437
12	2.179	3.055	4.318
13	2.160	3.012	4.221
14	2.145	2.977	4.141
15	2.132	2.947	4.073
16	2.120	2.921	4.015
17	2.110	2.898	3.965
18	2.101	2.878	3.922
19	2.093	2.861	3.883
20	2.086	2.845	3.850
21	2.080	2.831	3.819
22	2.074	2.819	3.792
23	2.069	2.807	3.768
24	2.064	2.797	3.745
25	2.060	2.787	3.725
26	2.056	2.779	3.707
27	2.052	2.771	3.690
28	2.048	2.763	3.674
29	2.045	2.756	3.659
∞	1.960	2.576	3.291

5. $V = 5.03$ L with a relative uncertainty $\Delta_V = 0.2\%$.
6. $V = 5.03(1 \pm 0.2\%)$ L.

3.5 ERROR PROPAGATION

Analytical instruments measure mass, volume, pressure, temperature, flow rate, concentration, etc. Every instrument has an uncertainty that contributes to the overall uncertainty of the variable. Even volumetric flow rate is a composite measure: we choose a time interval and then collect a volume of fluid in a graduated cylinder. Uncertainties include time (your response time, stopwatch precision), sampling the liquid, and instrument error. The scale must be precisely calibrated. Liquid volume depends on temperature, and even the volumetric flask is calibrated at a specified temperature.

We express the variable y as a function of the factors x:

$$y = f(x_1, x_2, \ldots, x_n).$$

The uncertainty of the variable (Δ_y) is the sum of the squares of the product of the uncertainties of each factor (Δ_x) and the partial differential of the function with respect to that factor:

$$\Delta_y^2 = \sum_{i=1}^{n} \left(\frac{\partial y}{\partial x_i} \Delta_i \right)^2.$$

This expression can be simplified for the case where the function y is a sum of factors or is a product of the factors raised to a power. For the sum of n factors:

$$y = \sum_{i=1}^{n} a_i x_i,$$

$$(\Delta_y)^2 = \sum_{i=1}^{n} (a_i \Delta_i)^2.$$

For a product of n factors,

$$y = \prod_{i=1}^{n} x_i^{a_i},$$

$$\left(\frac{\Delta_y}{y} \right)^2 = \sum_{i=1}^{n} \left(\frac{a_i \Delta_i}{x_i} \right)^2.$$

Example. You collect water in a flask for 10 s to calibrate a liquid mass flow controller (Table 3.7). Calculate the uncertainty of the mass of liquid, the time, and the mass flow rate.

Solution. The sample means and standard deviations are $\bar{m} = 48.4$ g, $s_m = 0.87$ g, $\bar{t} = 9.6$ s, $s_t = 0.22$ s, $\dot{m} = \bar{m}\bar{t}^{-1} = 5.03$ g s^{-1}, and $s_{\dot{m}} = 0.07$ g s^{-1}. $t(0.05, 6) = 2.447$, so $\Delta_m = 0.87$ g and $\Delta_t = 0.22$ s. To calculate the confidence interval for mass flow, we use

TABLE 3.7 Mass Flow Calibration

Test	m (g)	t (s)	$\dot{m}$ (g s^{-1})
1	48.2	9.7	4.97
2	48.5	9.4	5.16
3	47.0	9.4	5.00
4	48.4	9.6	5.04
5	49.5	10.0	4.95
6	49.4	9.8	5.04
7	47.8	9.5	5.03

$$\left(\frac{\Delta\dot{m}}{\dot{m}}\right)^2 = \left(\frac{a_1\Delta m}{\bar{m}}\right)^2 + \left(\frac{a_2\Delta_t}{\bar{t}}\right)^2,$$

where $a_1 = a_2 = 1$ because the mass flow rate is first order in mass and time. So

$$\Delta\dot{m} = \sqrt{\left(\frac{0.87}{48.4}\right)^2 + \left(\frac{0.22}{9.6}\right)^2} \cdot 5.03 \text{ g s}^{-1},$$

$$\Delta\dot{m} = 0.147 \text{ g s}^{-1} \approx 0.15 \text{ g s}^{-1}.$$

To be safe, express the uncertainty as 0.2 g s^{-1} or better as a percentage: $\Delta\dot{m} = 2.93\% \approx 3\%$. Finally, $\dot{m} = 5.0$ g s$^{-1} \pm 3\%$ (or $5.0(1 +/- 3\%)$ g s^{-1}).

3.6 REPORT DATA REALISTICALLY

Report both mean and uncertainty in your data. If you omit uncertainty, the number of significant figures conveys your measurement's certainty: The uncertainty (Δ_x) of 3.141 is ±0.0005. Carrying too many significant figures overstates the certainty in the data. Communicate your data precisely: as mentioned earlier, temperatures can vary by as much as $10\,°C$ to $20\,°C$ during experiments or across vessels; therefore, reporting five significant figures for temperature is excessive. Temperatures of $520\,°C$ and 793.15 K imply uncertainties of $\pm5\,°C$ and ±0.005 K, respectively, although the temperature is the same.

Fitted parameters to calculate specific heat, mass diffusivity, viscosity, etc., are reported with five or more significant figures. Atmospheric pressure is defined as 101.325 kPa, but barometric pressure varies by as much as 5000 Pa in 1 week (Patience, 2013). Derived variables such as conversion and selectivity depend on flow rates, species concentration, pressure, and temperature, each with its own uncertainty. At best, mass balances in complicated processes such as catalytic reactors are $\pm1\%$, but more typically the target is $\pm5\%$: a conversion, yield, or activation energy with three significant figures overstates the certainty.

To justify carrying three significant figures, how many times must you repeat the experiments? The sample size is calculated by solving for n in the relationship between the confidence interval and the standard error of the mean:

$$n = \left(z(\alpha)\frac{\sigma}{\Delta_x}\right)^2.$$

Example. Assume you have a population with a mean of 90 and a standard deviation of 2.

(a) What is the minimum sample size to achieve a confidence interval of 89.5 to 90.5 ($\Delta_x = \pm0.5$) at a confidence level (α) of 95%?

(b) If you purchase new instruments and increase the precision of the experiment, thereby reducing the standard deviation by a factor of 2, how many experiments do you have to perform to achieve $\Delta_x = \pm 0.5$?

Solution.

(a) Assuming that the sample size is greater than 30, we can assign $z(\alpha) = 1.96$:

$$n = \left(1.96\frac{2}{0.5}\right)^2 = 61.$$

(b) The number of experiments drops by a factor of 4 (since the sample size is proportional to the square of the standard deviation). Since the sample size is less than 30, we apply the t-statistic and have

$$n = \left(t(\alpha, n-1)\frac{\sigma}{\Delta_x}\right)^2.$$

You calculate n by trial and error using Table 3.6 to derive the t statistic. Alternatively, with our expression for the t-statistic as a function of n, you can solve for n explicitly:

$$n - \left[\left(-0.441 + \frac{2.40n}{n-1.01}\right)\frac{\sigma}{\Delta_x}\right]^2 = 0.$$

The value of n is 21, and $t(\alpha, n-1)$ is 2.08, which is six more experiments than if you assume a normal distribution with $z(\alpha) = 1.96$.

3.7 ERROR BARS

Nature (2014) requires that all graphs with errors bars state whether they are based on standard deviation (s), standard error of the mean (s_x), or confidence intervals (Δ_x). *Nature* adopts s.d., s.e.m., and CI to represent sample standard deviation, standard error of the mean, and confidence interval, respectively, whereas *Science* uses SD and SEM to represent sample standard deviation and standard error of the mean, respectively, As shown earlier, error propagation is based on confidence intervals, and thus we recommend that confidences intervals be represented by error bars on graphs and represent uncertainty of physical quantities (Δ_x, not sample standard deviation and standard error of the mean).

Confidence intervals are not yet the standard for error bars on graphs. In 2012, in 49% of all bar graphs and scatter plots in *Nature Methods*, the error bars expressed the standard error of the mean (Krzywinski and Altman, 2013a), in 45% they expressed the standard deviation, and in 5% their meaning was unspecified. In only one graph did the error bars express the CI based on a 95% confidence level.

The confidence interval is the product of the t-statistic (or z-statistic) at a defined confidence level α and the standard error of the mean. For a large sample size ($n > 30$), it is double the standard error of the mean for a 95% confidence level, α. The confidence interval is greater when $n < 30$ since the confidence level is based on the t-statistic, but even for $n = 6$ the confidence interval is only about 2.5 times the standard error of the mean.

For $n = 6$ the confidence interval is equal to the standard deviation; it is greater than the standard deviation for $n < 6$, and it is less than the standard deviation for $n > 6$.

3.8 EXERCISES

1. Correct the following expressions:
    a. $d_p = 2.5 \times 10^{-6}\,\mathrm{m}$
    b. $T = 793.15\,\mathrm{K}$
    c. $\$3962\,\mathrm{t}^{-1}$
    d. $Q = 1.46 \times 10^3\,\mathrm{m}^3\,\mathrm{s}^{-1}$
    e. $\Delta_Q = 6.9 \times 10^2\,\mathrm{m}^3\,\mathrm{s}^{-1}$
    f. $\Delta H_f = -1501.660\,\mathrm{kJ\,K}^{-1}$
    g. $E_a = 50.208 \pm 0.053\,\mathrm{kJ\,K}^{-1}$
    h. Ash content = 12.34%
    i. An estimate of 0.693
    j. An increase of 39% and 35.6%
    k. This correlation is accurate to within 3.2%:

$$S_r = 2.12(1 + 0.279S)^{0.156}St^{0.0755}Re^{0.377}.$$

2. Identify acceptable expressions:
    a. 7.928 ± 0.0495
    b. 7.928 ± 0.049
    c. 7.928 ± 0.05
    d. 7.93 ± 0.05
    e. 7.93

3. Correct the expressions that contravene SI or math writing conventions:
    a. ΔH_f
    b. $45 \pm 5\,\mathrm{m\,s}^{-1}$
    c. $45(5)\,\mathrm{m\,s}^{-1}$
    d. $(45 \pm 5)\,\mathrm{m\,s}^{-1}$
    e. Kelvin

4. Calculate the t-statistic for a sample size of 12 and a confidence interval of 99%. What is the difference between the value based on the correlation and the value reported in Table 3.6?

5. Calculate the average and standard deviation of the number of papers published for each discipline and $\phi_{25,0.1}$ from Table 1.6. Can you apply Chauvenet's criterion to any of the data?

Chapter 4

Graphs

Chapter Outline

After the abstract and conclusions, people look at images—pictures, schematics, charts, histograms, graphs—then tables, and finally text. Take time to make clear, unambiguous, and self-explanatory graphics because they contribute to people's first impression of your work (Kamat et al., 2014a). Graphs communicate the greatest amount of information in the least amount of space—they are more efficient than words. They display trends better than tables in less space. Tables archive data, which allows researchers to retrieve individual data points. Extracting data from x to y plots requires an additional step, but computer programs have made this task less arduous. For multiple related variables and response factors, tables are better. If the trends are simple (straight lines, fewer than four data points) or intuitively obvious, describe the data in the body of the article. Some journals recommend limiting the number of figures/schemes in an article to five (Kamat et al., 2014a).

Choosing the scaling length is the first step to make a graph: in papers, it is the length of the y-axis; in presentations, it is height of the slide (e.g., 190 mm in PowerPoint); in posters, it is the length or width of the poster, whichever is smaller. All dimensions in the graph—line weight and character font—depend on the scaling length. In papers, graphs are constrained by the width of the journal column: most images and tables must fit within a single column. Exceptionally, journals extend them across two columns or even the whole page. When graphic designers resize graphs to fit into the journal

Communicate Science Papers, Presentations, and Posters Effectively. http://dx.doi.org/10.1016/B978-0-12-801500-1.00004-8

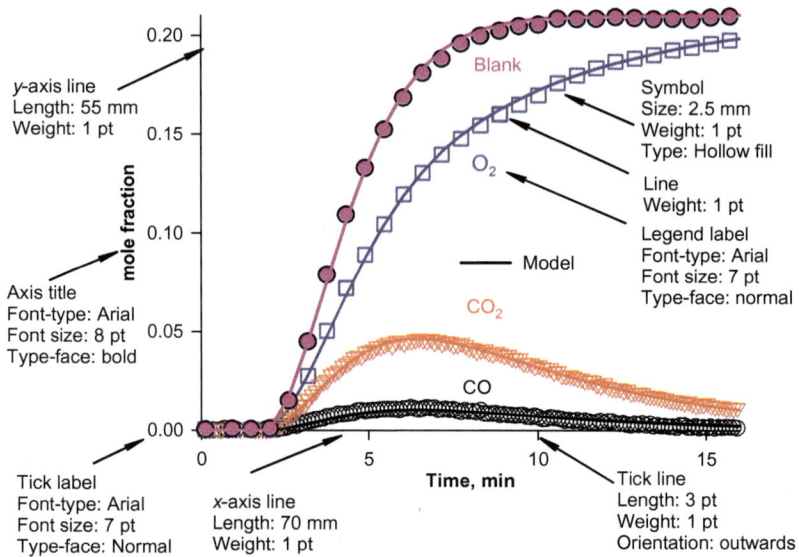

FIGURE 4.1   Standard dimensions for a figure 70 mm wide and 55 mm high.

columns, text becomes illegible. Templates from computer packages rarely provide camera-ready graphics. They are convenient to identify trends, but you have to modify them for the journal. We recommend making a template based on the graphs in the journal in which you wish to publish your work. The dimensions outlined in Figure 4.1 fit a column width of 85 mm, which is standard for journals with two columns per page.

The major elements of graphs are

1	Type	Scatter plot, box plot, pie chart
2	Dimensions	Journals, presentations, posters
3	Text	Typography, axis titles, labels, legends
4	Axes	Linear, logarithmic, scale breaks
5	Representing data	Symbol size, line weight, color
6	Lines	Ticks, grids, frame, confidence intervals
7	Titles and captions	Number of words, font size, typeface, position

The final composition—geometry of the plot, line weight, font type, symbol shape, and color—is a matter of taste, but in the paper they should all taste the same: apply a common template to scatter plots, bar charts, histograms, contour graphs, and three-dimensional (3D) plots. Line weights and symbol sizes should be heavier and larger in presentations compared to graphs in articles (Tables 4.1 and 4.2). Graphical elements in posters should be proportional to the dimensions of graphs in articles.

TABLE 4.1 Standard Dimensions (mm) of Plots in Papers, Presentations, and Posters

	Papers	Presentations	Posters
Plot Height, mm	55	55	150
Symbols	2.5	4	8
Text			
Axes titles	3	6	11
Tick labels	2.5	5	11
Legends	2.5	5	11
Lines			
Data	0.35	0.7	1
Axes	0.35	0.7	1
Tick length	1	2.1	3
Tick width	0.35	0.7	1

Note: For the same *y*-axis length, line weights and font sizes are double those in papers.

TABLE 4.2 This Table Reproduces the Same Data in Table 4.1 but Reports the Values in Points (pt)

	Papers	Presentations	Posters
Plot Height, pt	55	55	150
Symbols	7	12	24
Text			
Axes titles	8	16	32
Tick labels	7	14	28
Legends	7	14	28
Lines			
Data	1	2	3
Axes	1	2	3
Tick length	3	6	9
Tick width	1	2	3

4.1   GRAPH TYPES

Scatter/Line Plots

The best way to identify trends in data is to start with a scatter plot (Patience, 2013). Scatter plots and line plots represent causal relationships between independent variables (factors) on the abscissa (*x*-axis) and dependent variables (responses) on the ordinate (*y*-axis). If the data spans several orders of magnitude, consider selecting logarithmic axes. For example, the number

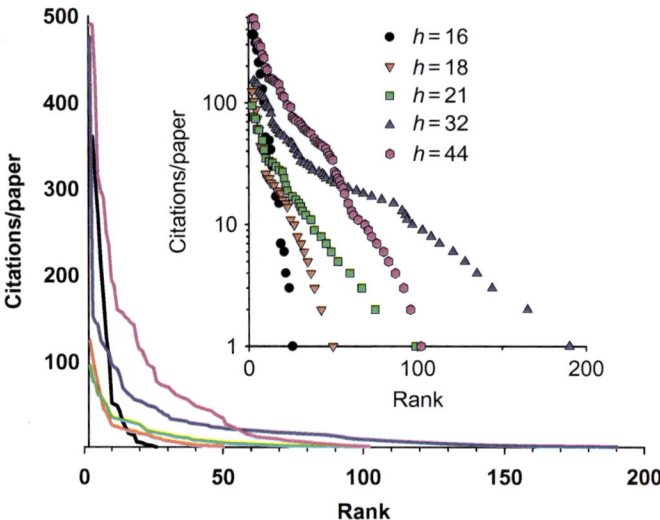

FIGURE 4.2 Citations per article versus article rank for five successful researchers with an h-index from 15 to 44. The colored lines in the plots correspond to the colored symbols in the insert. The relationship between article rank and citations per article follows a power law and is linear on a log-log plot (insert).

of citations of articles decreases with article rank following a power law (Figure 4.2): all authors have few oft-cited articles and many seldom-cited articles. Identifying the numerical values is easier with the log-normal plot (the insert in Figure 4.2). Modifying the axes to linearize the data is effective to establish the underlying relationship between the factors and responses.

Multiple scatter plots are graphs in which the response variables and factors are the same but the data represent variations in the experimental procedure or other factors. Figure 4.2 compares the number of citations per paper (response variable) versus paper rank (principal factor) for many authors (additional factors). The response variable in Figure 4.3 is maleic anhydride yield, and the principle factor is conversion. Yield and conversion depend on butane volume fraction, oxygen volume fraction (shown in Figure 4.3), and temperature, pressure, flow rate, and catalyst (absent from Figure 4.3). Multiple plots on the same graph can be confusing. Print graphs in black and white to ensure that symbols and lines are still identifiable—minimize symbol overlap.

Compare spectral data—infrared, X-ray diffraction—or analytical data from mass spectrometers, gas chromatographs, high-performance liquid chromatography instruments, etc., by placing one plot over another and offsetting the curve vertically (or horizontally) to avoid overlap (Figure 4.4). Identify peaks with numbers, stars, or other symbols on the graph, and describe what they represent in the caption or legend.

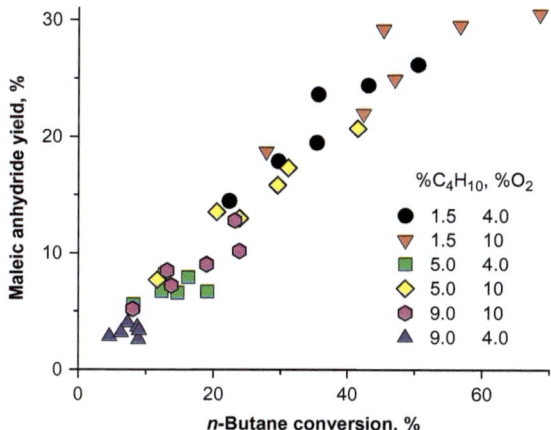

FIGURE 4.3 Choose symbol shape, color, and contour lines to differentiate data sets: maleic anhydride yield versus *n*-butane conversion.

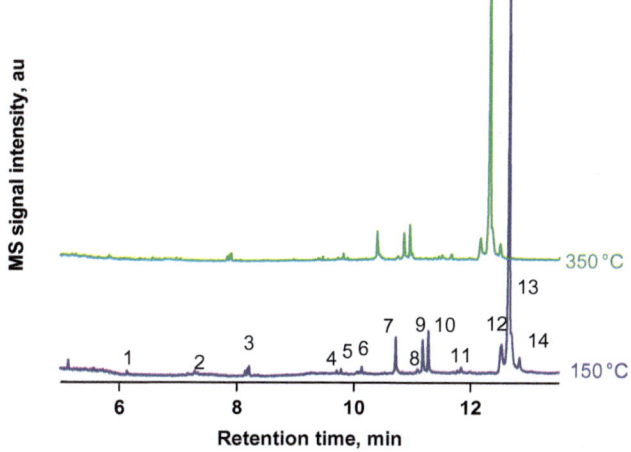

FIGURE 4.4 Mass spectrometer signal of compounds that were eluted from bitumen in 1 h at 150 °C (lower trace) and 350 °C (upper trace). The effluent stream passed through an aqueous quench. Toluene extracted the compounds that condensed in the quench. Peaks 1 through 14 are: ....

Bar Charts

Whereas scatter plots identify relationships between response variables and factors best, bar charts communicate categorical data best—groups of discrete data collected at some interval (days, months, years, age, country, experimental technique, reagent). In horizontal bar charts, we plot continuous data on the *y*-axis and the category on the *x*-axis. For example, we could plot the number

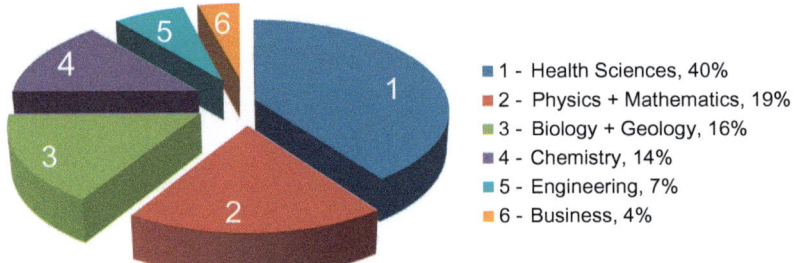

FIGURE 4.5   Proportion of journals assigned to each category defined by Google Scholar (2014).

of publications per country per capita (*y*-axis) versus country (*x*-axis, category). Pie charts are variations of bar charts. These diagrams are generally qualitative, but can be made quantitative by adding numerical values in the pies or in a legend (Figure 4.5). Marketing presentations rely on bar charts and pie charts to highlight differences between product performance or regional variations, for example.

Histograms are similar to bar charts, but plot quantitative data instead of categorical data. They represent the distribution of a response variable versus a factor. Often, the best way to represent this is with adjacent bars whose height corresponds to the number of observations (or the fractional frequency when the area is normalized to 1) in an interval you define. The most convenient bin interval equals the smallest measurable unit, or a multiple of this value. In 2013, sentences in the top 500 abstracts of *Science* articles averaged 20 words, with a standard deviation of 8 words (Figure 4.6). Of the 2700 sentences, one sentence had 65 words (maximum), one sentence had 3 words (minimum), and 169 sentences had 17 words (mode). The minimum bin size equals the range of values—62. Every bar (bin) represents the number of words in a sentence, and the height represents how many times a sentence had that many words. Increasing the bin size increases the number of observations per bin and smooths the curve.

Often a bar chart or histogram will have two or more variables or factors. For these cases, choose different colors for each variable. Identify each bin with a letter, abbreviation, or number to keep the histogram uncluttered, and add a legend to describe the terminology.

Three-Dimensional Plots

Scatter plots illustrate the relationship between multiple response variables and one factor (Figure 4.2). They also effectively demonstrate simple relationships between a single response variable and two or more factors (Figure 4.3). For complex relationships between a response variable and two factors, contour

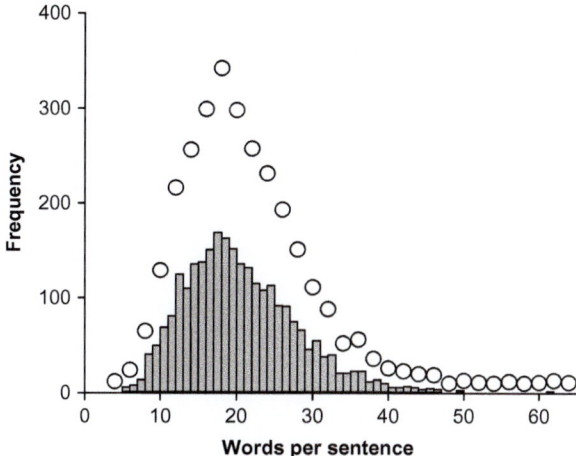

FIGURE 4.6 Histogram of the number of words per sentence in *Science*'s top 500 articles of 2013. The bin size for the bars increases by one word per sentence; the bin size for the open circles increases by two words per sentence.

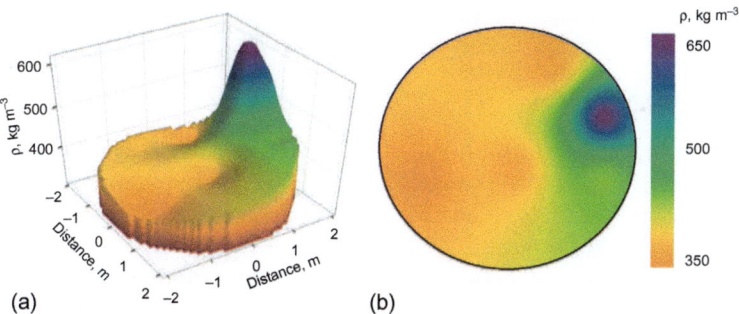

FIGURE 4.7 3D surface plots (a) and contour plots (b) of the solids suspension density across a 4.2 m diameter reactor. The high-density region exceeds $600 \, kg \, m^{-3}$ (purple), while the low-density region is at $350 \, kg \, m^{-3}$ (orange).

plots and 3D graphs are more practical. The factors could be spatial coordinates (Figure 4.7(a))—for example, time and a linear dimension, or temperature and pressure. Three-dimensional plots can be challenging to read, and retrieving data from them is difficult. Contour plots are clearer, but deriving quantitative data from them is also problematic. Shades of color are lost when articles are printed in black and white (Figure 4.7(b)). Grid lines help trace the contours, but having too many clutters the plot. A color legend demonstrates more clearly the variation of the response variable with the factors in the *x-y* plane.

Be careful when drawing the 3D mesh for your data. We generated the contour plot and 3D plot with only seven data points (Table 4.3). Computer

TABLE 4.3 Source Data to Construct the 3D and Contour Plots (Figure 4.7(a))

x-Coordinate (m)	y-Coordinate (m)	Suspension Density $(kg\,m^{-3})$
0.52	−1.55	630
1.41	−0.65	371
1.41	0.50	383
−0.52	−1.55	435
1.41	−0.65	357
−0.30	−0.10	368
−0.50	1.4	351

Note: Suspension density carries too many significant figures.

software generated the additional data to construct the image. Extrapolating beyond the data (to the edges of the vessel, for example), or even within the data, is dubious because the grid is so coarse. Figure 4.7(b) shows a gross mal-distribution of the solids suspension density. More experimental data are necessary to confirm the distribution, and repeated runs are important to understand the variance in the measurements.

4.2 DIMENSIONS

Publishers grant some flexibility to fit artwork, illustrations, and graphs onto the journal page. Most journals have either two or three columns per page. The page width is 170 mm and the column width is either 85 mm (two-column journals) or 58 mm (three-column journals). When you make your graph, set the length of the x-axis to 70 mm for two-column journals, or 50 mm for three-column journals. If the total width of the figure, including axis titles and legends, is greater than the journal column width, shorten the x-axis until it fits. If the legend doesn't fit in the plot frame, put it in the caption. Graphic designers may extend detailed images or graphs across two columns.

Design illustrations and graphs so that they fit into one column width, otherwise graphic designers may resize them. All text, symbols, and lines shrink together with the graph. The width necessarily includes all text outside the frame—axis titles, labels, and legends. An axis title, labels, and ticks represent 10% of the width of a graph. When a graph has two y-axes on either side of the plot, axis titles, labels, and tick represent as much as 20% of the width. Orient y-axis labels vertically rather than horizontally to save space. Also, choose a prefix for the units that minimizes the number of zeros before or after the period to reduce the number of characters:

| 0.0001 mm | 0.1 µm, or 100 nm |
| 1000 mg | 1 g |

Many journals prefer a square geometry for the graph frame when it is scaled to fit into one column. When the graph extends to two columns or more, the height is kept close to the one-column width. Keep these dimensions in mind when setting the line width and character heights. Scale characters and lines in a graph—font size, symbol size—with the plot's height, not its width. The width-to-height ratio of aesthetic graphs lies between the golden ratio (1.62:1) and a 1:1 ratio.

4.3   TEXT

Scaling text in graphs—axis titles, labels, legends, and figure captions—depends on the graphic designer as much as the original artwork. Be prepared to modify your plots on the galley proofs to ensure that the font is sufficiently large. Kamat et al. (2014a) insist that the minimum font size in graphs should be no less than half the font size of the text in the main body. We recommend that the font size in graphs be no more that 2 pt smaller than the caption font size. The size of the caption text should be equal to the size of text in the main body, or 1 pt less. For example, the text in *Science* articles is 9 pt Times New Roman. Helvetica and Arial are standard fonts for captions, but 8 pt Helvetica or Arial is as large as 9 pt Times New Roman. So, an Arial caption can either be 8 pt or 7 pt. Axis titles should be either the same font size as or 1 pt less than the caption text. All other text—tick labels, legends, symbol labels—should be 1 pt less than the axis titles (Tables 4.1 and 4.2).

Font Types

The geometric properties of text include font size, line height, and width. Other features to consider when sizing graph text are the font type (serif or sans serif) and the typeface (boldface, normal, or italic). *Serif* refers to the small projections at the extremities of some letters. *Sans serif* comes from the French word *sans*, which means *without*. Monospace characters are the same width, and can be serif or sans serif. The majority of newspapers and journals use serif fonts (e.g., Times New Roman) for the main body of the text (Table 4.4). Sans serif types are best for graphs, illustrations, presentations, posters, digitized media, and webpages—they are easier to read at low resolution. *Science* publishes the main text in Times New Roman, but all text in graphs and figure captions is in Helvetica. (Helvetica is common in Mac OS X and is similar to Arial in Windows). Terminal emulators, text editors, plain text e-mail messages, ASCII art, and URLs display text in a monospaced font.

TABLE 4.4 Character Constants for Serif, Sans Serif, and Monospaced Fonts (Pearson C., 2014)

Serif	η_f	η_{cap}	μ	Sans Serif	η_f	η_{cap}	μ
Baskerville	0.97	0.762	2.51	Arial	1	0.789	2.26
Georgia	1.09	0.713	2.27	Helvetica Neue	1	0.789	2.24
Palatino	1.10	0.688	2.26	Verdana	1.075	0.754	1.96
Times New Roman	1	0.740	2.48	Tahoma	1.075	0.754	2.25
`Courier New` (monospaced)	0.90	0.730	1.67	Trebuchet MS	1.046	0.757	2.2

Notes: η_{cap} is the ratio of the height of the uppercase letter to the font size; η_f is the ratio of the font size with respect to courier; and μ is the character constant, which is the ratio of the font size to the character width.

FIGURE 4.8   Times New Roman characteristic heights. The font size is the distance from the bottom of the descender to the top of the ascender. Roman numerals and uppercase letters are often the same height. Roman numerals in Times New Roman are slightly taller than uppercase letters. The baseline is an imaginary line on which letters and numbers sit.

Besides the character width (c_w), other features that change between font types are the character height and the ratio of the height of the uppercase letter (C_{lett}) to the font size (f). The font height is the distance from the bottom of the descender (the lowest point in letters that drop below the baseline—g, j, p, q, and y) to the top of the ascender (uppercase letters and roman numerals or the lowercase letters b, d, f, k, and l) (Figure 4.8).

For a given font size, the height of uppercase letters and the height of roman numerals in Arial and Helvetica are identical; they are taller than in Times New Roman. A 12 pt Times New Roman uppercase letter is the same height as an 11 pt Arial uppercase letter. The differences between the two fonts are more pronounced for lowercase letters: 11 pt Arial lowercase letters lie between 12 pt and 13 pt Times New Roman (Table 4.4). The ratio of the Courier font size to the Arial font size ($\eta_f = \frac{f_{cr}}{f_{arial}}$) is 0.9: for the same font size, Courier letters are 90% smaller than Arial or Helvetica letters. Lucida, Tahoma, and Verdana letters are 7.5% taller than Arial letters. Roman numerals and uppercase letters within the same font family extend from the baseline to the cap height, and are shorter than the font height by 68-79% (η_{cap} in Table 4.4). The capital letter Q is an exception for which the tail extends to the descender. Georgia's roman numerals are nonlining: 3, 4, 5, 7, and 9 extend below the baseline (1234567890).

Typography

Matching the text in graphs to journal standards requires a basic knowledge of typography. Images and text can be stored as either scalable vector graphics (SVG—vector format) or as individual pixels—raster graphics. A vector image can be zoomed into without limit and the quality remains the same. This is because vector formats store the instructions necessary to draw the final image. The program that displays the image has all the information it needs to view the image at higher resolution. Portable document format (PDF) is a vector format: when you zoom into a PDF image, the font will not become blurry. (PDF files can also contain raster graphics, in which case zooming in will blur the images.)

Characters and images in graphical abstracts are converted to raster graphics. In raster graphics, each pixel is stored as a value (with optional compression). When you zoom in, each pixel appears bigger, and eventually the image looks like stacked blocks. Antialiasing adds colored pixels around the blocks to improve the quality of the image. At high magnification, the blocks are visible; at low magnification, the colors make the image look curved.

We copied the number 9 (Arial, 11 pt) into Paint (Windows 7) and zoomed in to the maximum. Paint antialiases the number 9 with 12 colors—including black, dark red, orange, blue, and faint yellow (Figure 4.9(b))—rather than aliasing the character with black (Figure 4.9(a)). Compared with the vector image (Figure 4.9(c)), the image resolution at high magnification of the aliased and antialiased characters is poor.

The number 9 is six blocks by seven blocks. With an Arial 20 pt number 9, paint increases the grid to 11 blocks by 15 blocks, which improves the quality, but it remains unacceptable at the highest resolution; an Arial 48 pt number 9 (27 blocks × 35 blocks) has more black pixels than colored pixels, and is acceptable at the highest zoom.

Raster graphics express character density with the number of pixels. Font size and line widths are reported in points (pt) and ems (the width of the letter M) as well as millimeters (mm) and inches (in). Graphics programs and word processors generally size fonts and lines in points or millimeters. A 72 pt character is 1 in high. One pixel is 0.265 mm, so there are 3.78 px mm^{-1} (Table 4.5).

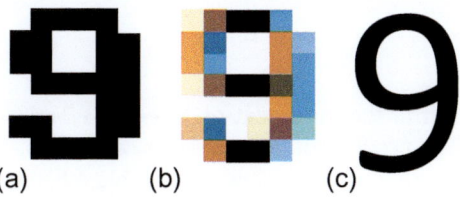

(a)   (b)   (c)

FIGURE 4.9   Raster and vector images of the number 9.

TABLE 4.5 Font Size and Line Conversion

point, pt	pixel, px	em	millimeter, mm	inch, in
0.5	0.7	0.04	0.18	0.007
0.75	1.0	0.06	0.26	0.010
1	1.3	0.08	0.35	0.014
6	8.0	0.50	2.1	0.083
7	9.3	0.58	2.5	0.097
8	10.7	0.67	2.8	0.111
9	12.0	0.75	3.2	0.125
10	13.3	0.83	3.5	0.139
12	16.0	1.00	4.2	0.17
14	18.7	1.17	4.9	0.19
16	21.3	1.33	5.6	0.22
20	26.7	1.67	7.1	0.28
24	32.0	2.00	8.5	0.33
36	48.0	3.00	12.7	0.50

We recommend that the font size of captions and axis labels be about the same as that of the text in the main body of the paper. Tick labels, legends, and other text in the plot should be no smaller than 1 pt less than the font size of the axis labels. At a minimum, all text should be greater than 6 pt. Since journals don't always specify the font size in graphs, you have to derive it from the journal font size.

Science publishes three columns per page, each 58.5 mm wide (w). Although vector images have no pixels, we use the nomenclature *pixels per character* to represent character width (c_w) and *pixels per line* to represent line density (λ). λ is the product of w and the conversion factor 3.78 px mm^{-1}: $\lambda = 221$ px line^{-1}. On average, *Science* has 50 characters per line (c_{pl}). The character width, c_w, is the ratio of the line density, λ, and the number of characters per line, c_{pl}.

$$c_w = \frac{\lambda}{c_{pl}} = \frac{58.5 \text{ mm line}^{-1} \times 3.78 \text{ px mm}^{-1}}{47 \text{ characters per line}} = 4.7 \text{ pixels per character.}$$

Every font type has a character constant μ that relates its size, f, to the character width (Table 4.4):

$$\mu = \frac{f}{c_w}.$$

The character font in the main text of *Science* is Times New Roman, for which $\mu = 2.48$. The product of its character width, c_w, and the character constant, μ, is 11.7 px (about 12 px), and from Table 4.5, the font size is 9 pt.

Axis Titles

Remember that figure and caption text is best in sans serif typefaces and that the character size of Arial/Helvetica is 1 pt bigger than that of some serif fonts such as Times New Roman. Axis titles should range from 7 pt to 8 pt Helvetica in *Science*, where the text font is 9 pt Times New Roman.

For graphs with vertical axis labels on both sides, the horizontal axis should be 60 mm wide. With only a single vertical axis label, the horizontal axis should be 70 mm wide. The overall width of the graph, including the *y*-axis title, label, and ticks, will be 85 mm, which fits in a column (Figure 4.1).

Abbreviate axis titles or write them explicitly, and place units after a comma or in parentheses:

- T, °C
- T (°C)
- T/°C
- Temp., °C
- Temp. (°C)
- Temperature, °C
- Temperature (°C)

The first construction is best for presentations because it has the fewest characters and thereby the most white space. The second-to-last construction is better for papers and posters because it's explicit. (Title formats and abbreviations are often dictated by the journal.)

Minimize the text within the plot (legends, callouts, explanatory text)—leave it for the captions. This recommendation applies to tick labels as well: for both axes, choose the unit prefixes that minimize the number of characters. For example, convert the axis labels from m to mm to report distances that range from 0.1 to 99 mm. Reporting 2 mm is better than reporting 2×10^{-3} m.

Avoid superfluous information in legends.

Temperature profile from simulation with $T_{\text{cooling}} = 690\,°\text{C}$ cooling temperature 690 °C

Temperature profile is redundant since the plot was temperature versus time. *Simulation* is unnecessary when you express data as symbols and models as lines. Substitute T for *temperature* and the subscript *cooling*.

Tips

- Minimize text in graphs, but not at the expense of clarity.
- Size your graphs for the journal—don't let the graphic designers shrink your artwork.

- Apply the same font size to captions and axis titles, and set all other text in the graphs to 1 pt less.
- Switch from serif fonts in the main body to sans serif fonts in figures.
- Expand the captions so the graphs are self-explanatory.
- Put the legends in the captions if they don't fit within the frame.

4.4   AXES

Scale the axes so that the experimental data extend across the entire plot—from bottom to top and from left to right. If the data range is far from zero, choose a value for the origin that is about 5-10% lower than the minimum value in the data. Ensure that the slope of the trend lines of the most important data is at an angle of 30° to 60° (Aken and Hosford, 2008). This spreads the data out so that people can read values from the graph more precisely.

For data that varies logarithmically, assign the nearest (lower) power of 10 of the smallest number in the data to the origin. For example, assign 100 to the origin when the lowest value is between 100 and 1000, and assign 0.0001 when it lies between 0.0001 and 0.001.

Rarely do axes warrant more than five major ticks (excluding the origin). Minor ticks are generally unnecessary when the scale is linear. If you choose to include them, put one or four minor ticks (corresponding to 0.5 and 0.2 intervals). For logarithmic axes, add one, four, or eight ticks between the major axes, which corresponds to 0.5, 0.2-0.4-0.6-0.8, or 0.2-0.3-0.4-0.5-0.6-0.7-0.8-0.9.

When graphs have a single abscissa and ordinate, don't frame the plot with axis lines on the top and to the right—maximize empty space. When a graph has multiple axes, axis lines on top and to the right are mandatory.

Breaks

In general, scale the y-axis to accommodate large peaks, and scale the x-axis to accommodate the range of the data. When the minor peaks are important, add another graph or expand the scale using a break function. Figure 4.4 fits the largest peak, but the other peaks too small. Adding a break at the second-highest peak heightens the minor peaks, and choosing a narrower x-axis range separates them (Figure 4.10). We split the graph in the middle and chose the scale so that the distance between each tick was about the same and the tick labels had the same magnitude—0.02. A plot with a logarithmic y-axis heightens the minor peaks with respect to the largest peak.

Inserts

When data occupies three quadrants of a plot, inserting images or an additional graph in the vacant quadrant is a good way to present the same data in a

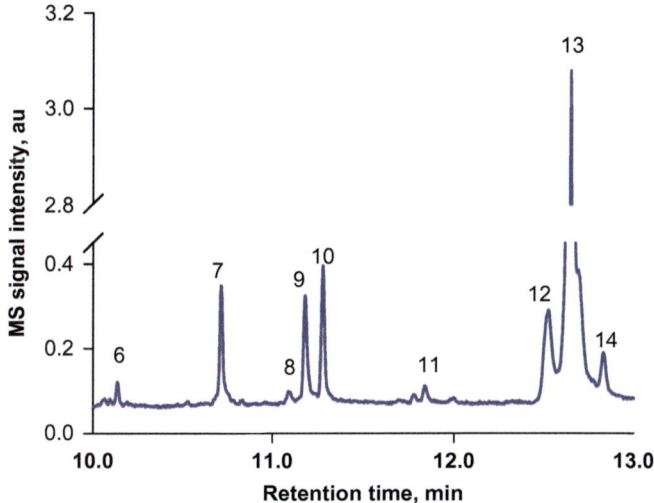

FIGURE 4.10 Plot breaks magnify small peaks of the mass spectrometer signal of compounds that were eluted from bitumen.

different way or a zoomed-in portion of the graph, etc. We reserve symbols for data and lines for graphs, but Figure 4.2 had so much data that we represented it with lines. The linear scales concentrated the data in a narrow area of the plot, so the symbols overlapped too much. This concentrated data leaves a large empty quadrant in the top right-hand corner, where we plot the same data but with a logarithmic y-axis. The log-normal plot spreads out the data better so that individual data points are visible for most of the quadrants.

Even with small symbols and the logarithmic scale, some symbols overlap in the top left quadrant (Figure 4.2). A log-log plot spreads out the data even further (Figure 4.11), which allows us to increase the symbol size by 30% to improve the visibility.

4.5 LINES

Lines that represent models or connect data are the same thickness as the axis lines. Ticks, error bars, grids, and arrows are thinner.

Graphics programs report line thickness (weight) in terms of millimeters, inches, points, or pixels. In articles, select 1 pt axis lines for plots that are 70 mm × 55 mm (x-y). For plots that are a couple of times smaller or larger, set the line weight proportional to the ratio of the length of the y-axis. Line weights are heavier for presentations. Double the line thickness for presentations with a 60 mm y-axis. The proportions between line weight and y-axis length in posters are the same as in papers. Since they are about three times larger, the axis lines

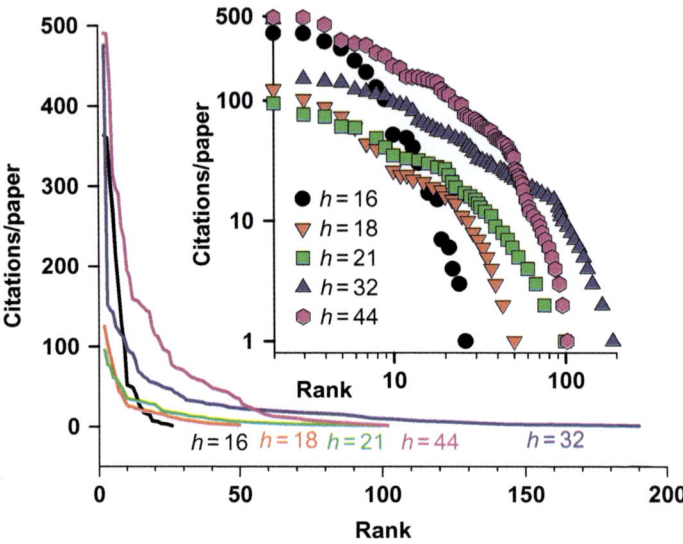

FIGURE 4.11 Author profile of citations per article versus rank and *h*-index.

are 3 pt. We recommend a line weight 1-2% of the length of the *y*-axis for papers and posters, and 2-4% for presentations (Table 4.1).

Ticks

We reiterate that minor ticks are unnecessary, except for logarithmic scales. The ticks should point outward so that they don't interfere with data and should extend three times the thickness of the axis lines. For a standard 70 mm × 55 mm plot, we recommend an axis line weight of 1 pt (0.35 mm) and a major tick length of 1 mm. The tick width is the same thickness as the axis lines. Minor ticks are half the length and half the weight of major ticks.

Grids

Avoid grid lines in the plot area. For log-log plots, set the line weight equal to half the line weight of the axes. Alternatively, make the lines gray or set the line transparency between 15% and 45% (Krzywinski, 2013). Graphs help the reader understand significant trends. Extraneous lines clutter figures. If readers require precise values from a plot, they can extract them from a PDF image with computer discretization technology.

4.6   REPRESENTING DATA

Reserve symbols for experimental data and lines for models and correlations. Symbols should include error bars. However, the size of the symbol can also represent the error. When graphs contain several sets of data, each set should have its own color and symbol type (circle, square, diamond, etc.) (Figure 4.3). Colors are useful even when the article is printed in black and white because photocopies convert color into *50 shades of gray*. Moreover, articles are available on the Internet in color. Lewandowsky and Spence (1989) demonstrated that color was the most effective discriminating feature in terms of pattern recognition. Letters of the alphabet are also effective symbols.

German psychologists developed visual perception theories in the 1920s to account for how people organize objects into groups. People assign greater importance to thick black lines than to thin gray lines. Similarly, people attribute greater importance to larger symbols than to smaller symbols or to a character shape that is distinctive: a cross is more distinctive than hollow squares, circles, and triangles.

Krzywinski and Wong (2013) stressed that symbol types should look alike with respect to size and complexity: symbols that are significantly different may bias readers and, as a result, they might attribute more or less importance to that series of measurements. Alternating hollow symbols and filled symbols helps discriminate between sets of data. Including a black contour around the symbol is important for light colors such as yellow, light green, and pink. We recommend an edge line thickness 10% of the symbol size. Figure 4.3 is 70 mm 55 mm, and the symbol size is 2.5 mm.

Figure 4.3 reports fewer than five sets of independent measurements with a total of 36 data points. We recommend that the symbol height be the same as the height of the characters on the *y*-axis. Set it to 2.5-5% of the length of the *y*-axis: graphs that are 50 mm high will have 2.5 mm symbols.

We drew the triangles in Figure 4.3 with a hairline edge and the squares with a 0.1 mm thick line. These lines are barely evident in the figure. The diamonds and hexagons have edge lines that are 10% of the symbol size. For light colors, a contour line is mandatory.

When there are hundreds of data points, shrink the symbols or alternate between filled and hollow symbols. Figure 4.2 has 6 data sets with 244 data points. With that much data, lines would have been more appropriate.

Many authors include trend lines. We discourage connecting dots or drawing straight lines or curves through the data. Rather, we recommend that you develop a phenomenological model to represent the data or regress it with a simple function and then add the regression line. This can help identify trends and interpret what they mean. When glycerol vapor reacts over a catalyst, coke forms on the surface and builds up with time (Figure 4.12). As much coke accumulates

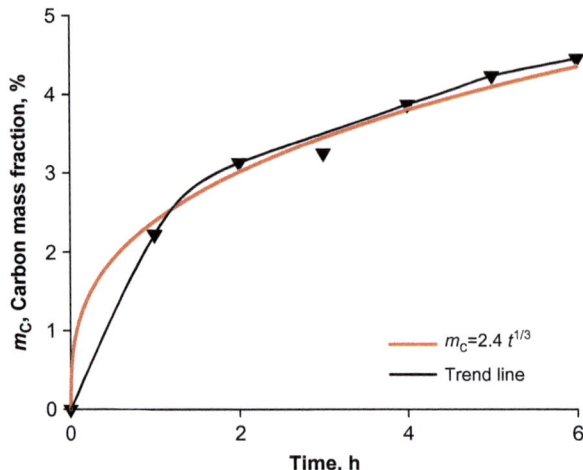

FIGURE 4.12 Glycerol reacts to form acrolein, but a fraction forms coke and this builds up with time.

on the surface during the first hour as builds up in the following five hours. We connected the data points, and fit the data to a one-third power law that accounts for 99% of the variance. The power law model infers that coke will continue to accumulate after 6 h but at a slower rate, and that half of the buildup in the first hour happened in a couple of minutes.

When there are multiple independent data sets in a single graph, symbol shape and color are insufficient to follow trends. For these cases, it is even more important to add lines, but draw regression lines based on a model.

The weight of lines representing a model should be approximately the same as the weight of the axis line. Trend lines should be less conspicuous—half the weight of the axis line, for example. State explicitly in the caption if the lines in the plot are trend lines (Kamat et al., 2014a).

4.7 TITLES AND CAPTIONS

Aken and Hosford (2008) place titles above tables and captions below figures. Rather than repeating the axis titles as the caption, describe the trends and what the data represents. *Science* and *Nature* allow authors to describe the figures with several hundred words—even more text than the abstract. Since many people might look only at the figures, they have to be self-explanatory. The caption must have enough information to guide the reader through the graphics, and it might comment on the trends.

The labels for captions and titles may be in bold or in a different color. Captions in *Science* are written in 8 pt Helvetica.

Don't repeat captions in the body of the paper. You can highlight trends in the graphs and discuss the underlying phenomenology. Avoid sentences that merely direct the reader to the figure or table or repeat the data without some comment on its significance or on the physical phenomena taking place:

Figure 1 shows the reaction rate as a function of temperature.	Increasing temperature from 120 °C to 140 °C doubled the reaction rate (Figure 1).

The sentence on the left is unnecessary. The statement on the right is explicit; it points the reader's attention to the graph and highlights the relationship between temperature and reaction rate.

Nature rejects metadiscourse/signposting in the main text of the journal to introduce tables and figures (left column below).

Figure 4.3 shows the relationship between butane conversion and maleic anhydride yield as a function of butane and oxygen concentrations.	Maleic anhydride yield increases linearly with conversion regardless of butane or oxygen mole fraction (Figure 4.3).

Saying that Figure 4.3 shows the relationship is redundant (metadiscourse). Add a comment concerning the trends in the graph, and attempt to relate them to a physical phenomenon. The sentence on the right-hand side is better.

4.8 EXERCISES

1. How would you improve the axis titles in Figure 4.10?
2. Identify four deficiencies in Figure 4.2.
3. Recommend a legend for the lines in Figure 4.2.
4. What is an appropriate line weight for a 50 mm wide plot for a paper?
   a. 0.1 mm
   b. 0.5 pt
   c. 0.5 mm
   d. 1.5 mm
   e. 1.4 pt
5. What is the minimum line weight for a 170 mm wide graph in a presentation?
   a. 1.7 mm
   b. 1.7 pt
   c. 3.4 mm
   d. 3.4 pt
   e. 10 pt
6. In a plot with hundreds of experimental data points in a single data set, how can you superimpose a line representing a model so that the symbols don't overlap excessively?
7. When your data occupies most of the quadrants of the plot, where do you put a legend for an article, a presentation, and a poster?
8. Is it ever appropriate to connect the experimental data (symbols) with lines?

9. *Science* publishes text in the main body of its articles in 9 pt Times New Roman. What is the optimum font for legends in figures?
   a. Arial 7 pt to 9 pt
   b. Arial 6 pt to 8 pt
   c. Helvetica 6 pt to 8 pt
   d. Times New Roman 8 pt to 10 pt
   e. Times New Roman 7 pt to 9 pt
10. For the same font size, are bold characters wider or taller (or both) than normal typeface?
11. *Nature Methods* publishes the main body of its articles in 9 pt Times New Roman.
    a. Identify the caption text.
    b. Text in the figures is in Arial/Helvetica. What is the minimum recommended font size for the text labels?
    c. Does *Nature Methods* respect the recommended minimum?
12. What is the recommended thickness of contour lines in a 2D plot, or grid lines in a 3D plot?
13. Figure 4.13 (Winzer et al., 2012) shows one panel of a 13-panel graph dedicated to characterizing virus-induced gene slicing. Enumerate five deficiencies of the plot.
14. Are multisentence captions appropriate for a presentation? Are they appropriate for a poster?
15. How can you best present multiple data sets in a single graph in a presentation?
16. If text in Times New Roman is 75 mm wide, how wide would it be in Arial?
17. Is it ever appropriate to include a graph drawn by a software package of analytical equipment manufacturers in a presentation, paper, or poster (Figure 4.14)? How can you improve the figure?
18. How would you improve Figure 4.15?
19. Enumerate at least 10 deficiencies of Figure 4.16.

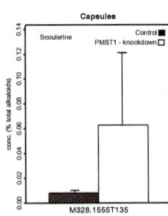

FIGURE 4.13 Winzer et al. (2012) published a 6 panel plot to demonstrate virus induced gene silencing. This plot includes a control with error bars.

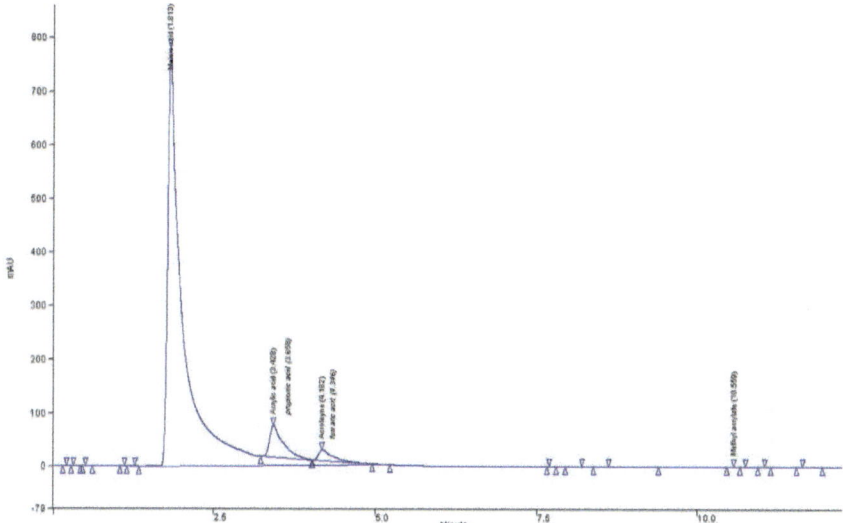

FIGURE 4.14   Standard high-performance liquid chromatography graph drawn by the software package from the analytical instrument manufacturer.

FIGURE 4.15   Scanning electron microscope image of calcined vanadium pyrophosphate.

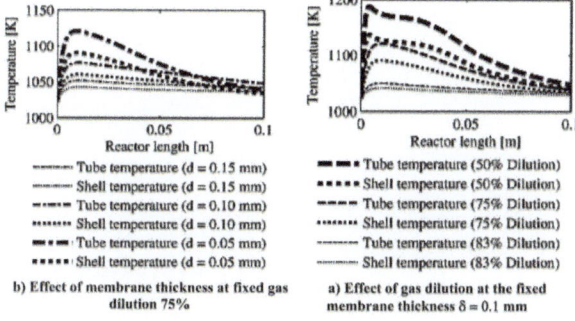

FIGURE 4.16   Methane oxidative coupling. (From Godini et al. (2012).)

Chapter 5

Tables

Chapter Outline

Science publishes five times more figures than tables (Table 5.1). Each figure averages five photos, schematics, and graphs—which makes 25 times more images than tables in the printed journal. The supplementary materials on the website contain more figures and tables, ones that *Science* excludes from the printed journal. On the website, figures outnumber tables by a 2:1 ratio. So, even though it is not published in the journal, the data are available for those who need it.

Tables assemble classes of information, including descriptive data (text), property data (text and parameter values), numerical values, surveys, government forms, images, and equations. Descriptive tables compare products, compare performance criteria, summarize literature, and serve to organize complex lists. The cells contain text, images, and equations, and may contain numbers.

Property tables list the physical characteristics of systems, processes, and materials. They contain a descriptive column, one or more columns of values, and optionally a column of units and references.

Data tables contain multiple columns of numerical values, for which short column heads are sufficient to describe what they are. Journals are less likely to publish tables of raw data, and prefer graphs.

Davis (1997) recommends that tables contain at least six data points. When the list includes unrelated elements, a table with three data points is acceptable. Choose the orientation that minimizes space but generally orient like elements vertically. If you have data in a table, don't repeat it in the text (Day and Gastel, 2011), rather highlight the trends, identify the significance of the values, or discuss exceptional values.

In Table 5.1, the last line lists the totals of the columns. We didn't go through every single data point in the table when describing it. Instead, we highlighted

TABLE 5.1 *Science* Publishes More Graphs and Images than Tables

Issue	No.	Tables	Figures	Images	Supplementary Tables	Supplementary Figures
6089						
Reviews	1	0	3	13	0	0
Articles	0	0	0	0	0	0
Reports	13	5	43	208	48	143
6090						
Reviews	1	0	2	3	2	22
Articles	3	0	13	52	10	42
Reports	12	5	45	202	49	75
6106						
Reviews	1	0	2	20	0	0
Articles	1	0	5	18	26	5
Reports	13	19	44	300	37	113
6111						
Reviews	1	0	4	8	0	0
Articles	2	1	9	65	31	79
Reports	13	13	40	249	18	91
Total	61	43	210	1138	221	570

our conclusions so readers will understand what we want to convey (they might have their own interpretation, but at least they will understand what we thought).

If you can describe data with a couple of sentences in the text, a table is unnecessary. When there are multiple relationships and a combination of descriptive data and numerical values, tables help organize the information efficiently—meaning it takes less space and less time for the reader to understand it. Furthermore, they illustrate the relationships between factors and variables to make them obvious.

You must mention all your tables in the text. Don't write *Table # shows…* This phrase is signposting (metadiscourse) and redundant, especially if you restate the caption. Rather, describe Table # or its context, and at the end of the sentence add (Table #).

Favour graphs to tables when you present experimental data. Tables are appropriate for lists or when there is little data. Write explicit and informative titles. Avoid sentences. State the variable name or its symbol in column titles. Abbreviate the titles—*T* instead of *Temp.* or *Temperature*—unless you have no nomenclature section. Include the units in the title so you needn't report them after each value in the table. Most data warrant no more than three significant figures; two is better. All numbers in a column have the same number of significant figures. Numbers that straddle multiples of 10 are exceptional: 10.3

has three significant figures, whereas 9.8 has only two. In this example, rather than the number of significant figures, the digit is significant.

Presentations and Posters

Carrying more significant figures is justified for a large number of experiments, but for most presentations it draws attention away from the message; you needlessly clutter the slide. More white space on a table is better than reams of numbers: eye tests are for optometrists.

In some cases, the audience might recognize a table format with 50 or more data points—an economic analysis, lists of components in samples, chromato-graphic results, etc. You aren't drawing attention to the individual data points, rather you are displaying a construct that they relate to. If you want to talk about certain numerical values in the table, increase the font size of those values to more than 18 pt. State the data and allow the audience to believe what you say. A scientific presentation is rarely an examination. Discuss detailed tables after the presentation with the more interested or doubtful.

5.1   TERMINOLOGY

Captions

Captions go above tables and start with the label: the word *Table* (optionally capitalized), the table number, and a separator. The separator could be a period, a colon, an em dash, a new line, or just a space. The whole label can be bold. Examples include the following:

Table 1.1:
Table 1.1.
Table 1.1 —
TABLE 1.1

Journals dictate the label style.

The caption starts with the title of the table, which can be bold. *Science* adds text after the title to describe the table in more detail. When the table includes uncertainties, the caption specifies whether the uncertainty refers to standard deviation, standard error, confidence intervals, or instrument resolution.

In captions that are several sentences long, you can expand acronyms, list equations, describe the context, the date, and the duration of the experiments, and detail experimental characteristics and anomalies. Anything relevant to the table can go in the caption. Some journals require only a title.

This caption lists the acronyms in the table (which we have omitted):

Table 3. Common tumor biomarkers that can potentially be used as nanoparti-cle targets. PSMA, prostate-specific membrane antigen; PCLA, prostate cancer lipid antigen; MUC1, mucin-1; VCAM-1, vascular cell adhesion molecule 1;

VEGFR, vascular endothelial growth factor receptor; Tem1, tumor endothelial marker 1; TAMs, tumor-associated macrophages; TAFs-FAP, tumor-associated fibroblasts-fibroblast activation protein; TEMs, Tie-2-expressing monocytes; APA, aminopeptidase A.

Cheng et al. (2012, p. 906)

The font type in the caption is often the same as that in the table, which may be sans serif (e.g., in *Science*; see Chapter 4).

Minimize text in presentations' table titles. Remember that the table is a complement to your voice. Mention the abbreviations and other significant facts about the table while you speak.

Posters are more like papers. The captions must be self-explanatory, and you might need even more text to make the table clear, since a poster doesn't have the benefit of 3000 words to support and explain the data. Presentations don't need captions. Tables don't need labels (e.g., Table #) in presentations or in posters.

Headings and Cells

Headings include (Table 5.2) spanner headings, subspanner headings, stub headings, row headings, and column headings. The stub is the first column. Spanners extend over several columns. Lines above the columns identify what spanner headings apply to. Spanner headings are always centered. Column headings and cells are centered if the column contains numerical data or short text. Column headings and cells are left-aligned if the column cells contain long text. Stub headings and row headings are often left-aligned, unless they are numerical, in which case they're centered.

Though some people make their headings bold, don't do this. The line separating the heading section from the rest of the table is enough to emphasize the headings. Often the journal dictates the conventions about bold headings.

TABLE 5.2 Table Terminology (Rabinowitz and Vogel, 2009; Davis, 1997)

	Spanner Heading[a]			
	Subspanner Heading 1[b]		Subspanner Heading 2	
Stub Heading	Column Heading 1	Column Heading 2	Column Heading 3	Column Heading 4
Row heading 1	Cell 1[c]	Cell 3	Cell 5	Cell 7
Row heading 2	Cell 2	Cell 4	Cell 6	Cell 8

[a]Footnote a (e.g., source of data).
[b]Footnote b (e.g., explanation of heading).
[c]Footnote c (e.g., qualifying statement).

Cells containing numerical data should be aligned at the decimal point if the orders of magnitude of the data are close to each other. Otherwise, either center them or right-align them.

Choose column headings so they don't extend over the width of the cells. If they do, add appropriate subscripts to variable symbols or abbreviate the heading, and describe the variable or abbreviation in the caption or footnote. For example, instead of writing *Optimal growth temp* (Day and Gastel, 2011, p. 93), write $T_{\text{optimal growth}}$. If your paper has a nomenclature section, you don't need to restate the abbreviations in the caption. In the worst case, split the heading over at most two lines.

Many journals add extra space between columns until the table width equals the journal column width. Because they do so means you have more flexibility in naming your headings. However, respect the format of your other tables and keep your headings short.

Scale units such that the corresponding values are between 0.1 and 100 (Chapter 3). For example, instead of writing 0.01 ms and 1000 μs, write 10 μs and 1 ms, respectively. Values between 0.1 and 100 take less space. Use scientific or, preferably, engineering notation when (a) the orders of magnitude of the values in a column differ by more than 10^3; and (b) you cannot change the prefix (we generally don't add prefixes to temperature units) and the values are greater than or equal to 10^5 or smaller than 0.001.

The SI recommends that column headers be the variable divided by its unit, for example, t/s (BIPM, 2006). This makes mathematical sense because the column heading is equal to the value in the cell, so $t/s = 1 \Rightarrow t = 1\,s$. It is unambiguous to add powers of 10 to the unit in the column heading if you wish to do so instead of changing the unit prefix. To have multiples of 10^7 s, write $t/(10^7\,s)$ in the column heading. As before, $t/(10^7\,s) = 1 \Rightarrow t = 1 \times 10^7\,s$.

A flaw with this convention is that column headings will be wider than the data in cells, especially if the units are long. Although this isn't a problem for tables with few columns, some tables will be too wide for the journal. Instead of using this convention for some tables and putting the unit below the variable in the heading in other tables, follow the latter convention for all your tables.

Footnotes

Footnotes address information relevant to specific elements of the table—for example, anomalous data, different operating conditions, the source of the data, explanatory notes, and qualifying statements (Rabinowitz and Vogel, 2009). When the information applies to many rows or columns, put it in the caption.

Footnote symbols can be numbers, letters, or glyphs (e.g., asterisk, dagger, double dagger). Because numbers can be confused with exponents, favor letters and glyphs (but not both in the same table).

Assign footnote symbols sequentially from left to right, starting at the top and going down. Start the sequence of footnote labels from the beginning for every table. Continue footnotes on the same line as the previous ones to save space.

Lines

Excluding spanner lines, use three horizontal lines at most: one at the top of the table, one to delimit the heading section, and one at the bottom. *Science* omits the top line. The top and bottom lines should have the same weight. They may be heavier than middle lines, which may be heavier than spanner lines.

Don't shade rows, and don't separate them with lines. You don't need to emphasize the separation between rows of data or single-line cells. For multiline cells, consider increasing the white space between each row.

Never use vertical lines (Fear, 2005). If you want to separate two sections of a table, use spanners. *Science*, *Nature*, and other high-impact journals don't use vertical lines.

Never use double lines (Fear, 2005), be they horizontal or vertical. The more lines you have, the more cluttered the table looks. White space is better.

5.2   DESCRIPTIVE

Descriptive tables are lists of associated elements such as *name*, *composition*, and *target* (Table 5.3). These tables are composed mostly of words, but may contain numerical values. You can describe associated elements in the text of your paper, or with a list, but even for only two elements, a table is better.

Table 5.3 lists nanoparticle attributes. The advantage of this kind of table is that you can easily identify what you're looking for without having to sift through dense text. Most of all, it's an efficient use of space because it directly highlights important features of the elements of the table.

The stub heading is *Name* and the row headings are the names of the nanoparticles. The column headings don't extend over the width of the columns' cells. We can improve the table by renaming the stub *Nanoparticle* and left-adjusting the first column heading.

Table 5.4 is a descriptive table that breaks many of our rules:

- The heading section is shaded.
- The headers are bold.
- There are too many horizontal lines.
- There are vertical lines.
- *Date* is left-aligned and *Duration* is centered, which is inconsistent.
- The heading of the fourth column needs two lines.

We modified Table 5.4 to respect our guidelines (Table 5.5): The table has more white space, and the rows and columns are distinct even without the

TABLE 5.3 Targeted Nanoparticles in Clinical Development for Use in Cancer Therapy

Name	Composition	Target	Comments	Status
MCC-465	PEGylated liposomal doxorubicin containing antibody GAH targeting agent	Tumor antigen	Metastatic stomach cancer	Phase 1
MBP-426	Liposomal oxaliplatin containing human transferrin protein targeting agent	Transferrin receptor	Advanced or metastatic solid tumors	Phase 1/2
SGT-53	Liposomal p53 cDNA containing antibody fragment targeting agent	Transferrin receptor	Solid tumors	Phase 1
CALAA-01	siRNA-loaded polymeric nanoparticles containing human transferrin protein targeting agent	Transferrin receptor	Solid tumors	Phase 1
BIND-014	Docetaxel-loaded polymeric nanoparticles containing peptide targeting agent	Prostate specific antigen	Advanced or metastatic cancer	Phase 1

Source: From Cheng et al. (2012, p. 905).

TABLE 5.4 Absolute Dates in Geologic Time

Unit	Symbol		Number of Years Ago
	Date	Duration	
kiloannum (*no hyphen*)	ka	ky	10^3
mega-annum	Ma	My	10^6
giga-annum	Ga	Gy	10^9

Source: From Rabinowitz and Vogel (2009, p. 434).

TABLE 5.5 Absolute Dates in Geologic Time (Revised)

Unit	Symbol		Years
	Date	Duration	
kiloannum	ka	ky	10^3
mega-annum	Ma	My	10^6
giga-annum	Ga	Gy	10^9

TABLE 5.6 Top Ten Unproductive Author Responses

	Author's Response to Editor	Editor's Reaction to Author's Response
1	The reviewer selected by the editor to review our paper is not an expert.	The reviewer is sometimes the one that was suggested as a preferred reviewer.
2	The editor chose a wrong reviewer. This reviewer has a strong bias towards our work.	The reviewers are selected from a general pool of physical chemists and chemical physicists. Editors attempt to avoid reviewers with obvious conflicts of interest, either pro or con. Furthermore, authors are encouraged in their submission cover letter to inform editors of any potential conflicts with researchers in the field.
3	The reviewer misunderstood our experiments/results.	If the reviewer misunderstood the results, the author needs to explain the results more clearly. Revising the text or presenting the results in a different format may help resolve the misunderstanding.
4	The reviewer is wrong, and their comment does not deserve an explanation.	This does not provide any useful information in terms of why the reviewer is wrong or mistaken. Explain in detailed scientific terms what is incorrect.
...	...	...
10	I am not a native English speaker. You should not expect me to write well.	An effective and grammatically correct presentation is required since reviewers cannot comprehend and, therefore, adequately evaluate poorly written and/or poorly composed papers. Papers published in an English language journal must be written in proper English. Authors can seek assistance from language editing services or native English speakers to help address language difficulties.

Source: From Kamat et al. (2014b, p. 898).

horizontal and vertical lines. We vertically centered the headings of the unit and years columns.

Table 5.6 is another descriptive table that doesn't respect our guidelines:

- The font in the first two columns is red.
- There are too many horizontal lines.
- There are vertical lines and a colored double vertical line.

Table 5.7 respects our guidelines: The font is black everywhere, there are three horizontal lines and no vertical lines. Since the rightmost column contains

TABLE 5.7 Top Ten Unproductive Author Responses (Revised)

	Author's Response to Editor	Editor's Reaction to Author's Response
1	The reviewer selected by the editor to review our paper is not an expert.	The reviewer is sometimes the one who was suggested as a preferred reviewer.
2	The editor chose a wrong reviewer. This reviewer has a strong bias toward our work.	The reviewers are selected from a general pool of physical chemists and chemical physicists. Editors attempt to avoid reviewers with obvious conflicts of interest, either pro or con. Furthermore, authors are encouraged in their submission cover letter to inform editors of any potential conflicts with researchers in the field.
3	The reviewer misunderstood our experiments/results.	If the reviewer misunderstood the results, the author needs to explain the results more clearly. Revising the text or presenting the results in a different format may help resolve the misunderstanding.
4	The reviewer is wrong, and the reviewer's comment does not deserve an explanation.	This does not provide any useful information in terms of why the reviewer is wrong or mistaken. Explain in detailed scientific terms what is incorrect.
...	...	...
10	I am not a native English speaker. You should not expect me to write well.	An effective and grammatically correct presentation is required since reviewers cannot comprehend and, therefore, adequately evaluate poorly written and/or poorly composed papers. Papers published in an English language journal must be written in proper English. Authors can seek assistance from language editing services or native English speakers to help address language difficulties.

multiline cells, we put more white space between the rows than usual. Otherwise, that column would look like a block of text.

5.3 PROPERTY

Property tables are for data in which the stub elements have no continuous mathematical relationship. They contain more numerical values than descriptive tables and more text than data tables. The row headings consist of descriptions or variable names. The rest of the cells contain numerical values or units.

Property tables include lists of material properties and reagents, nomenclature equations, and comparisons of experimental conditions between different authors.

In property tables, most rows contain data that are independent of the data in the other rows. Inlet pressure is independent of capital depreciation, which is independent of the cost of oxygen (Table 5.9). The variables in each row have different units, but they're all process design variables. It does not make sense to reproduce these variables either in a bar chart or in an *x-y* plot because they are independent of each other: *x-y* plots imply a causal relationship between response variables and factors.

When variables have the same unit, you can either represent them as a bar chart or put them in a property table (Table 5.8).

The footnotes to Table 5.8 refer to text in the caption and in the table. This table doesn't respect the guideline of Davis (1997) that tables should contain six or more elements. However, listing the information in the text would be unwieldy because the different elements in the stub would blend into the text.

This is what the table looks like as text:

"Average bulk chemicals are worth 1000 $/t biomass, transportation fuels are 200 $/t biomass to 400 $/t biomass, and generating electricity is 60 $/t biomass to 150 $/t biomass (*48*). The values are based on costs in the Netherlands, but the order of the values is likely to be similar across the developed world. Cattle feed is worth 70 $/t biomass to 200 $/t biomass (*58*). This range of values depends on the quality of the feed. Landfilling waste costs −400 $/t biomass."

The biomass uses and their values are more difficult to extract from the text than from the table. Furthermore, the table lists the elements in decreasing order of value, which is immediately obvious.

TABLE 5.8 Approximate Valorization of Biomass Waste for Different Uses[a] (*48, 58*)

	Value ($/t Biomass)
Average bulk chemical	1000
Transportation fuel	200-400
Cattle feed[b]	70-200
Generating electricity	60-150
	Cost
Landfill	−400

Source: From Tuck et al. (2012, p. 695).
[a]Taken from (*48*) apart from data for cattle feed. The values are based on costs in the Netherlands, but the order of the values is likely to be similar across the developed world.
[b]Data from (*58*); this range of values depends on the quality of the feed.

TABLE 5.9 Problem Data

Data	Value	Unit
Ethane feed	50	kt/y
Inlet pressure of the system	16	bar
Pressure loss per unit	0.5	bar
Capital depreciation	3	y
Annual operation	8000	h
Cost of ethane	72	$/t
Cost of oxygen	50	$/t
Price of acetic acid	730	$/t

Source: Adapted from Montolio-Rodriguez et al. (2007, p. 5604).

TABLE 5.10 Thermal Properties of the Materials Used in the Gallium Arsenide Furnace

Material	Symbol	Conductivity ($W\,m^{-1}\,K^{-1}$)	Emissivity
Solid gallium arsenide	Ga As S	7.2	0.36
Liquid gallium arsenide	Ga As L	17.1	0.36
Graphite	G	42.0	0.64
Quartz	Q	3.0	0.5
Steel	S	27.2	0.45
Boric oxide	B	1.7	0.5
Graphite felt	F	1.0	–

Source: From Dupret et al. (1990, p. 1867).

An interesting way to shorten your methods section is to put design conditions, economic costs, and operating parameters in a table (Table 5.9). As with Table 5.8, you save on words because you don't need to construct sentences to contain the information. Also, the information is easier to see because it's not hidden within the text.

The units are in a separate column (Table 5.9). Though they could have been in the row heading, placed in parentheses after the term, they are more distinguishable in a separate column. They wouldn't be aligned if they were in the row headings, so they would be harder to see. Furthermore, the rows read as sentences. The first row translates to *The ethane feed was 50 kt/y*.

Thus far, all property tables contained numerical values or units in the columns. Table 5.10 contains a column of text: the *Symbol* column. Putting a column of units is impractical because each column has a different unit (or none).

It would be ambiguous which column the unit applies to; a unit in a row could apply to all values in the row.

A column of data (with the same unit) in a table contains three things: the variable, the unit, and the cells. For each cell, the column satisfies the following mathematical equation: variable = cell value × unit. This is similar to the SI convention except that the unit goes in a separate row to narrow the table columns.

- If you can add a prefix to the unit such that the cell value is between 0.1 and 999, do so. For $L = 5 \times 10^{-9}$ m, the variable is L, the unit is nm, and the cell value is 5.
- If you cannot add a prefix to the unit because the data are too big or too small, divide the variable by a power of 10 that brings the cell value to between 0.1 and 999. For $E = 5 \times 10^{-17}$ J, the variable is $E/10^{-17}$, the unit is J, and the cell value is 5.
- If the variable is nondimensional, divide it by a power of 10 that brings the cell value to between 0.1 and 999, and leave the unit field empty. For $N_{Re} = 1.2 \times 10^{6}$, the variable is $N_{Re}/10^{6}$, the unit is empty, and the cell value is 1.2.

To respect our guidelines, the *Material*, *Symbol*, and *Emissivity* headings should be one row higher (Table 5.10). This makes it clear that *Material*, *Symbol*, and *Emissivity* have no unit, and makes the parentheses around the unit of *Conductivity* unnecessary.

5.4   DATA

Table 5.11 contains seven factors and five response variables. To demonstrate relationships between the variables and the factors would take multiple plots in multiple graphs. Summarizing the data in a table saves space. Researchers can more easily manipulate the data, make their own plots and graphs, and compare their data quantitatively; they don't have to discretize a graph to extract precise values.

Most readers uninterested in modeling prefer graphs to data tables. Data tables may be better off in the supplementary materials on the journal's website.

Had Table 5.11 followed the SI convention of dividing the variable by the unit, it wouldn't fit on the page. In fact, the original table has more columns.

The table must specify whether the mole fractions and MAC refer to the inlet or outlet conditions. Adding subscripts expressing this information to the factors would increase the table width; it's also repetitive. Rather than adding subscripts to these factors, we created the spanners In and Out. This saves space and is not repetitive.

TABLE 5.11 Reactor Performance Data in the Pilot Plant

	In			Out							
T °C	y_{C_4} %	y_{O_2} %	y_{O_2} %	MAC kg h^{-1}	y_{CO_2} %	y_{CO} %	X %	S %	τ_{regen} min	τ_{riser} min	M_s t h^{-1}
379	9.4	9.2	1	38	5.1	4.5	33	67	2.9	0.56	30
381	10.1	5.2	0.9	43	4.3	4.2	22	72	1.8	0.29	47
371	9.2	5.1	0.8	35	3.4	3.2	23	74	2.4	0.4	37
370	4.9	3.5	0.8	28	1.5	1.5	31	69	2.5	0.42	35
377	6.8	6.7	1.3	38	3.7	4.6	35	70	2.2	0.38	40
376	12.9	10.9	0.9	41	4.6	4.3	28	67	2.9	0.56	30
373	5.9	3.8	1.1	32	1.7	1.8	23	64	2	0.31	42
380	6.4	4.9	1.2	37	2	2.3	25	61	2	0.3	42
383	11.8	12.9	3	52	5.7	5.7	31	70	2.4	0.37	38
378	11.6	13	1	46	5.5	5.4	33	68	1.8	0.54	29
379	11.5	12.7	1.2	43	5	5	34	67	1.9	0.55	28
374	13.3	14	2.4	48	5.4	5.1	32	66	4.2	0.5	33
370	8.8	11.1	3.3	39	4.9	5.1	34	65	4.4	0.51	28
371	8.7	8.9	2	34	3.5	3.3	31	70	2.9	0.48	30
381	16.3	16.8	4.1	40	5.9	4.7	23	64	1	0.6	24
387	16.2	15.1	4.1	36	5	4.1	21	64	0.8	0.59	22
387	12.8	12.9	3.7	34	4.9	4.4	25	65	1	0.57	23
387	17.3	13.2	3.5	40	4.2	3.6	19	66	0.7	0.44	28
384	11.9	14.9	3.7	42	5	4.6	32	65	1	0.65	21
383	14.8	16.2	3.8	43	6.1	5.8	26	63	0.9	0.67	20
383	15.8	14.7	3.1	39	5.9	5.2	22	64	0.9	0.69	20

Source: From Patience and Bockrath (2010, p. 6).

The position of the last digit in each number indicates the uncertainty in the data. Several cells in this table carry an incorrect number of significant figures. For instance, in the fourth column, the 1, 2, and 3 should be reported as 1.0, 2.0, and 3.0.

All the numbers are aligned at the decimal point, which makes the data easier to read. The units are such that all values are between 0.1 and 999.

When a table's values are similar to each other, the table is unnecessary (Table 5.12). Peng et al. (2012) state, "The O_2 consumption on discharge follows the cell current (Fig. 3A), and the charge-to-mass ratio is $2e^-/O_2$ on each discharge (Table 1)." Their table (Table 5.12) reports charge-discharge cycles from 1 to 100. They mention in the text that all the values are about 2, but they ignore the third significant figure of the ratios. On the one hand, if the third significant figure were significant, they should have said so in the text: Including the table in the article leads the reader to believe that the difference in the numerical values is important. Simply mentioning the table in the text

TABLE 5.12 Ratios of the Number of Electrons to Oxygen Molecules on Reduction (Discharge) and Oxidation (Charge)

Cycle Number	Discharge e^-/O_2	Charge e^-/O_2
1	2.01	1.98
5	1.99	2.04
10	2.02	1.98
100	2.03	2.01

Source: From Peng et al. (2012, p. 564).

is insufficient to explain it. On the other hand, if the third figure were insignificant, the table would be redundant because they already mentioned that the value is 2.

5.5   EXERCISES

1. Improve Table 5.13.
2. Becke (2014) reviewed 50 years of density-functional theory (DFT). To prove the significance of the field, he listed the number of times the top six articles were indexed in Web of Science and Google Scholar—Table 5.14. Improve the table.
3. Improve Table 5.15.
4. Improve Table 5.16.
5. Improve Table 5.17. In the table the Royal Society of Chemistry published, the *Surface area* heading is on one line, and the word *temperature* in the

TABLE 5.13 Comparison Between Velocity Perturbations and Stefan Velocity for Two Growth Conditions, $L = 10\,cm$

	Case 1	Case 2
C_0' $(m\,s^{-1})$	130	130
P' (Torr)	0.1	1
D' $(cm^2\,s^{-1})$	120	15.4
U'_{Stefan} $(\mu m\,s^{-1})$	4.86×10^3	0.440
U' $(\mu m\,s^{-1})$: interface	0.835	0.282
bulk	7.5	7.5

Source: From Zappoli (1990, p. 1835).

TABLE 5.14 Total Citations for Notable DFT Papers as of January 1, 2014

Paper	Web of Science	Google Scholar
LYP	43 123	49 703
B3PW91	42 642	52 028
PBE	30 575	37 771
B88	24 766	28 529
KS	21 670	31 251
HK	15 222	27 317

Note: References Are as Follows: LYP,[67] B3PW91,[97] PBE,[59] B88,[63] KS,[4] HK.[3]

TABLE 5.15 Estimated Rate Constants (Revised)

k_d (s^{-1}) = 9.592×10^{15} exp[-1.343×10^5 (kJ/kmol)/RT (K)]
k_p (m^3/(s kmol)) = 1.310×10^{14} exp[-7.065×10^4 (kJ/kmol)/RT (K)]
k_t (m^3/(s kmol)) = 8.498×10^{21} exp[-8.851×10^4 (kJ/kmol)/RT (K)]
k_{tm} (m^3/(s kmol)) = 2.886×10^1 exp[-4.268×10^4 (kJ/kmol)/RT (K)]
k_{ts} (m^3/(s kmol)) = 4.487×10^6 exp[-6.816×10^4 (kJ/kmol)/RT (K)]
$k_t = k_{tc} = k_{td}$
$k_{tcd} = k_{tc}/k_{td} = 6.629 \times 10^{-11}$ exp[-6.125×10^4 (kJ/kmol)/RT (K)]

Source: From Chang and Liao (1999, p. 150).

TABLE 5.16 Bounds on Node Pressures

Node	Pressure Lower Bound (bar)	Pressure Upper Bound (bar)	Node	Pressure Lower Bound (bar)	Pressure Upper Bound (bar)
29	40	86	105	40	56.8
30	40	86	110	40	67
62	40	49	114	85	85
66	40	86	119	40	81
76	40	81	141	60	86
98	40	81	144	40	56.8
99	61	86	Other	40	68.7

Source: From Tabkhi et al. (2010, p. 957).

TABLE 5.17 Amberlyst 46 Characteristics

Catalyst	Surface Area $(m^2\,g^{-1})$	Average Pore Diameter (Å)	Total Pore Volume $(mL\,g^{-1})$	Acidity (meq.) $(H^+$ per g)	Max Working Temperature (°C)
Amberlyst 46	75	235	15	>0.4	120

Source: Adapted from Pirola et al. (2014, p. 46923).

TABLE 5.18 Optimized Kinetic Parameters for FFA Esterification Using Different Models

Model	SSE	$k_1{}^0$ $(mol*s^{-1}*m^{-3})$	$k_{-1}{}^0$ $(mol*s^{-1}*m^{-3})$	Ea_1 $(kJ*mol^{-1})$	Ea_{-1} $(kJ*mol^{-1})$
Pseudo-homogeneous (IDEAL)	0.131	$9.30*10^7$	$9.62*10^{-8}$	32.268	-67.589
Pseudo-homogeneous (UNIQUAC)	0.137	$2.12*10^8$	$1.13*10^{-7}$	33.154	-64.343
Adsorption-based (IDEAL)	0.145	$9.67*10^6$	$8.39*10^{-9}$	43.152	-57.229
Adsorption-based (UNIQUAC)	0.291	$1.34*10^7$	$1.09*10^{-10}$	43.067	-65.658

Source: From Galli et al. (2014, pp. 982-983).

Max working temperature heading is on the top line. We adapted it to fit our column width.

6. Improve Table 5.18.
7. Improve the legends in Figure 4.16.

Chapter 6

Paper Essentials

Chapter Outline

Journals require you to conform to a standard structure, and many adopt the IMRAD format—introduction, methods, results, and discussion. Most require an abstract, a list of references, and an acknowledgments statement. Conclusions are sometimes included as part of the discussion section, but generally journals require a separate section. *Nature* and *Science* adopt a newspaper/magazine format: they don't explicitly divide the paper into predefined sections. The flow of information is essentially the same except that the methods and conclusions sections are absent in *Science*.

Don't wait until you have compiled all the data to start writing. The very act of writing inspires new ideas, new ways of looking at the research, and (worst of all) new experiments. Write while you do the work. As soon as you've assembled most of the data, order the figures and tables to make a coherent story. This approach saves time, and even helps identify areas that require more detail or data.

Professional science writers take 8 h to produce 500 words; researchers might take 50% more time. Making a graph takes up to 4 h, and making a table takes 1 h or more. Writing a 3000 word manuscript with six graphs and two tables could take about 2 weeks. Editing, proofreading, and revising adds more time. Pinker (2014) revises chapters twice before asking for feedback, twice after that, and twice before submitting his books to editors: six revisions to get it right.

Communicate Science Papers, Presentations, and Posters Effectively. http://dx.doi.org/10.1016/B978-0-12-801500-1.00006-1

Submitting the manuscript on the journal's website can be frustrating (meaning you allotted less time than it takes). We estimate that organizing the manuscript, writing text, preparing graphs, and revising the manuscript many times takes 80 h to 100 h excluding the research. If you practice writing, it will take less time.

Structure your paper with mind maps (Gelb, 1988). This technique relies on free association in which your mind wanders from one idea to another. You draw pictures representing these ideas and connect them with lines. Add keywords to the pictures to help write the text. You can use this technique to organize the paper, write sections and paragraphs, and even attack sentences.

After you have ordered the graphs, add sentences beside each of them to explain the significant trends or highlight anomalies and discoveries. At this point, review how you ordered them with your coauthors: It is easier to move a few plots than it is to move paragraphs and pages of text. After you've collected all the data, be critical and decide what is appropriate to report and what is unnecessary.

During the first draft, concentrate on writing and documenting your ideas rather than correcting spelling errors as you type. Correct the spelling and grammar later. Try a pen instead of a text editor or word processor. Typing can be slower because we have the tendency to correct spelling mistakes and grammar while composing text.

- write;
- correct;
- type;
- correct; and
- revise and revise again.

Write a lead first (Burger Associates Inc., 1991), not necessarily the introduction. Try the 15 s test: state every important element of your paper in 15 s. Then write an outline. Then write as if you are explaining it to someone (out loud). Then edit the text. Finally, fix your content. So lead, outline, write, edit, revise, and revise.

6.1   TITLE

The title represents the body of the work. It contains the subject and at times the objective and even the results. It should have the minimum number of words that represent the content (Day and Gastel, 2011). Craft the title to suit the audience and the journal. Assertive, descriptive, enumerative, and comparative titles are four common types.

In 2013, *Science's* 500 most cited articles and *Nature's* 500 most cited articles together averaged 10.5 words per title, with a standard deviation of 2.5 (Figure 6.1). None of the titles were longer than 19 words; only one was 19 words long, and nine were 17 or 18 words long. So, limit the number of the

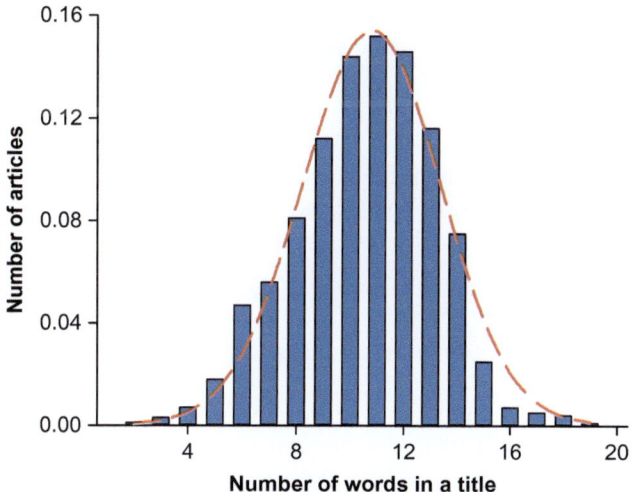

FIGURE 6.1   The 1000 most cited articles in *Nature* and *Science* (2013) averaged 10.5 words per title. Only 10 titles had more than 16 words. A normal distribution accounts for 98% of the variance in the data.

words in your title to 15 or fewer and closer to 10. Prusiner (1998) published a paper entitled "Prions"—a one-word title—in the *Proceedings of the National Academy of Sciences of the United States of America*, and it was among the top 1000 articles between 1989 and 2014. Although we suggest that short titles are desirable, limiting the number of words in the title might also reduce the likelihood of people finding your paper.

Assertive Titles

Assertive titles are sentences (Silyn-Roberts, 2013), and occur most commonly in journalism. Here are excellent examples from the *JAMA*:

- Nationwide IV Fluid Shortage Threatens Care (Kuehn, 2014)
- Regulation Boosts Vaccination Rates (Voelker, 2014)
- Cancer Care Shows Signs of Strain as Patients Live Longer (Mike, 2014)
- Stroke Risk May Be Increased After Shingles Episodes (Jin, 2014)

These titles are accessible even to nonpractitioners. The subject and results are easy to grasp. But not everyone agrees that assertive titles are appropriate: Rosner (1990) regards titles as labels, not sentences. He maintains that assertive titles trivialize the scientific process and relegate it to a product, an answer, or an acceptable outcome. The authors might nuance their conclusions more in the body of the paper, and this is lost altogether with an assertive title. Jin (2014) hedged their title with *may be increased*.

We hold that active verbs communicate data, ideas, methods, and conclusions more powerfully than passive verbs, so write active assertive titles.

Descriptive Titles

Descriptive titles include the subject, the outcome, or the means—the experimental technique, process, or specialized analytical equipment. These are the most common title formats. They are labels, not sentences:

- Ultrathin oxide films and interfaces for electronics and spintronics (Bibes et al., 2011)
- Memory effects in complex materials and nanoscale systems (Pershin and Ventra, 2011)
- Metalloproteins Containing Cytochrome, Iron-Sulfur, or Copper Redox Centers (Liu and K. Ai, 2014)
- Chemicals from Alkynes with Palladium Catalysts (Chinchilla and Nájera, 2014)

Descriptive	Assertive
Kinetics of mixed copper-iron based oxygen carriers for hydrogen production by chemical looping water splitting Chiron and Patience (2012, p. 10526)	Cu-Fe mixed oxides split water

The descriptive title above contains the subject (kinetics), the end product (hydrogen), the feed (water), the type of catalyst (mixed copper-iron oxide), and the technology (chemical looping)—many keywords, which is good. However, people skimming through titles of this journal might miss the essence of the work, which is splitting water. The assertive title is compelling because it focuses on the most interesting aspect of the work: splitting water. The technology is less important than what the catalyst does.

Enumerative Titles

Enumerative titles include lists of applications, techniques, compounds, etc. The lists easily become unnecessarily long. A hanging title separates the label from the list of items with a colon.

- Identification, Characterization, and Natural Selection of Mutations Driving Airborne Transmission of A/H5N1 Virus (Linster et al., 2014)
- Recovery of a solvolytic lignin: Effects of spent liquor acid volume ratio, acid concentration and temperature (Thring et al., 1990)
- Biochars prepared from anaerobic digestion residue, palm bark, and eucalyptus for adsorption of cationic methylene blue dye: Characterization, equilibrium, and kinetic studies (Sun et al., 2013)

The last item in the list above has 22 words and 154 characters. It could be stated more clearly with fewer words. We suggest titles should contain 15 words or less and fewer than 100 characters.

Comparative Titles

Web of Science (WoS) has 870 000 comparative titles that include *versus*, *vs.*, and *compared with*. This represents 2.2% of the 39 000 000 titles.

- Rivaroxaban versus Warfarin in Nonvalvular Atrial Fibrillation (Patel et al., 2011)
- Trastuzumab in combination with chemotherapy versus chemotherapy alone for treatment of HER2-positive advanced gastric or gastro-oesophageal junction cancer (ToGA): a phase 3, open-label, randomised controlled trial (Bang et al., 2010)
- Apparent Versus True Carrier Multiplication Yields in Semiconductor Nanocrystals (McGuire et al., 2010)
- Troponin elevation in coronary vs. noncoronary disease (Agewall et al., 2011)
- Land use transitions: Socio-ecological feedback versus socioeconomic change (Lambin and Meyfroidt, 2010)

Enumerative (hanging) titles are often too long.

Gold Nanoparticles: Assembly, Supramolecular Chemistry, Quantum-Size-Related Properties, and Applications toward Biology, Catalysis, and Nanotechnology Daniel and Astur (2004, p. 293)	Gold Nanoparticles in Biology, Catalysis, and Nanotechnology: Assembly, Supramolecular Chemistry and Quantum Properties
Semiconductor quantum dots and related systems: Electronic, optical, luminescence and related properties of low dimensional systems Yoffe (2001, p. 1)	Semiconductor quantum dots: Electronic, optical, and luminescent properties
Polydopamine and Its Derivative Materials: Synthesis and Promising Applications in Energy, Environmental, and Biomedical Fields Liu and K. Ai (2014, p. 5057)	Polydopamine and Its Derivatives: Synthesis and Energy, Environmental and Biomedical Applications
One-dimensional Titanium Dioxide Nanomaterials: Nanowires, Nanorods and Nanotubes Wang et al. (2014b, p. 9346)	TiO_2 Nanowires, Nanorods and Nanotubes

In the last example, we removed the word *nanomaterials* and left the nanomaterials. *One-dimensional* is redundant because wires, rods, and tubes are one-dimensional shapes. Alternatively, the title could read *One-dimensional Titanium Dioxide Nanomaterials*, but search engines will target words such as *nanowires* and *nanorods* better than *one-dimensional*.

Abbreviations, Acronyms

We sampled 140 articles from *Chemical Reviews* (impact factor 41.3), *Annual Review of Neuroscience* (impact factor 20.6), *Cell* (impact factor 32.0), *Advances in Physics* (impact factor 34.3), *Nature* (impact factor 38.0), *Science* (impact factor 31.0), and *IEEE Automatic Control* (impact factor 2.7). The abbreviations in the titles of this sample were:

TH17, p53, RNA, $Li-O_2$, Mfsd2a, BRAF, CRISPR/cas9, AcrAB–TolC, YidC, Gabaa, AMPA, fos, jun, miRNA, siRNA, MTOR, RNP, CD1d, IL-10, k-10, SN 2013cu, LQ, H_2, A.C., JAK2, polyADP, and mtDNA.

Many of these abbreviations demonstrate the curse of knowledge (Pinker, 2014). The authors know what they are talking about, but unless you are a specialist in the field, they don't mean anything to you. Granted, the rest of the title does put the abbreviation into context, but should you abbreviate anything in a title? Aken and Hosford (2008) say that acronyms are fine if you first define them and then use them more than three times. What about a chemical formula? WoS reproduces chemical formulas, as do other bibliometric databases, but some convert formulas to HTML, which makes them illegible: hydrogen sulfide (H_2S) becomes H₂S, and titanium dioxide (TiO_2) becomes TiO₂.

Many titles refer to analytical instruments. Unless your research topic is analytical methods and instrumentation, don't refer to them in the title. If your subject is instrumentation, then it is pertinent to put the name in the title and maybe even the abbreviation.

WoS indexes 5487 articles from 1989 to 2014 with *Fourier transform infrared* in the title; 16 340 articles abbreviate the words with *FT-IR, FTIR,* or *FT IR*. The proportion of these forms remained constant from 1989 to 2014. Stating analytical techniques can introduce zero words such as *use of, using, study of,* and *investigation*:

Analysis of Different Types of Soil by FTIR and ICP-MS Zhao et al. (2014, p. 3401)	FTIR and ICP-MS Spectra of Montmorillonite Farmland Soils

Spectroscopy and Spectral Analysis published this article, so abbreviations are acceptable, but the first five words are superfluous. Rather, the authors could have dedicated more words to farmland soils and the results. Lai et al. (2014) state their title succinctly while avoiding zero words: *Raman and FTIR spectra of CeO_2 and Gd_2O_3 in iron phosphate glasses.*

TABLE 6.1 Superfluous (Zero) Words in Titles (Chapter 2)

Word	WoS		Nature, Science	
	$N/10^3$	%	N	%
Studies	1853	4.8	3762	2.6
Use	1834	4.7	283	0.2
Analysis	1288	3.3	173	0.1
New	993	2.6	1012	0.7
Application	595	1.5	105	0.1
Development	592	1.5	561	0.4
Characterization	505	1.3	1102	0.8
Novel	448	1.2	151	0.1
Performance	428	1.1	161	0.1
Assessment	301	0.8	197	0.1
Investigation	246	0.6	109	0.1
Enhanced	231	0.6	333	0.2
Advances	170	0.4	1166	0.8
Observation	124	0.3	154	0.1
Modification	78	0.2	642	0.4
Total	9686	24.9	9911	6.9

Notes: Each row represents the number, *N*, and percentage of titles that contain the word. In total, WoS numbers 39 000 000 articles and Nature and Science together number 144 000 articles.

Activate Your Title

Assertive titles are concise and vigorous. Make titles that transmit the most information with the least number of characters. Maximize keywords. Eliminate descriptive nouns that are implicit to all articles—studies, development, investigation, etc. (Table 6.1). *New* and *novel* are superfluous adjectives. Two-thirds of all titles in WoS Core Collection contain the suffix *ion* or *ing*, which is not necessarily verb mutilation, but we expect that a significant fraction of these titles could be better. As many as 25%—close to 10 000 000—of the titles contain zero words. They are three times more frequent than in *Nature* and *Science*. The top four, *studies, use, analysis*, and *new*, appear in 15% of the titles in all the journals. The lower frequency of these words in *Nature* and *Science* supports our contention that they are irrelevant.

Tips for Titles

- Keep the title between 8 and 15 words long.
- Avoid words such as *studies, use, analysis, new, novel*, and *innovative*.
- Minimize nouns derived from verbs.
- Activate the title.
- Focus the title on the results, subject, and novelty, but not the means.

- Spell out any acronym you didn't learn in high school: RNA and DNA are appropriate.
- Write the names of elements—gold, nickel, iron—when they constitute the subject of the research.
- Abbreviate the names of inorganic compounds and catalysts: $Pt/BaO/Al_2O_3$.
- Abbreviate the names of simple and common molecules: $NaCl$, NO, HCN.
- When the name of a simple/common molecule has roughly the same number of syllables as the molecular formula, write it out: *water* is preferred to H_2O (unless it is part of a molecular formula).
- In some cases write out the molecular formula as well as the name of the compound: bichromium(III)-substituted tungstosilicate (Wang et al., 2014a); $[N(CH_3)_4]_4[A\text{-}\beta_{(12)}\text{-}SiW_{10}Cr_2O_{36}(OH)_2(H_2O)_2] \cdot 5\,H_2O$.
- Don't abbreviate the names of organic molecules.

6.2 ABSTRACT

After the title, the abstract used to be the second most important part of a manuscript (Rabinowitz and Vogel, 2009). Graphical abstracts and research highlights are now read more than abstracts. Since these two constructs are derived from abstracts, writing abstracts well is still important. Like for a title, a short abstract is better than a long rambling treatise, but when a journal specifies a range of 150-300 words, aim for 300. Summarize the major contributions such that the reader appreciates the significance of the work without having to read the entire document. Focus on the results, not the means.

Write the abstract several times: at the beginning (even before you've collected all the data), when the paper is nearly complete, and at the end to reflect the finished manuscript. Organizing and writing the abstract brings clarity. As with all writing, do it rapidly. Preparing a mind map for the overall abstract and even for the first sentence can also be helpful (Gelb, 1988). Minimize passive sentences and identify the agents. When you discuss specific phenomena you measured/derived/observed as part of your study, use the past tense. However, keep the present tense to discuss what the results imply.

The *New England Journal of Medicine* formalizes the abstract with headings. Most journals are less explicit, but we recommend you follow the *New England Journal of Medicine*'s structure: Dedicate the first one or two sentences to the background. In the next couple of sentences, present the method. After that, discuss the results and how they are important or significant. Finally, close the abstract with a few concluding remarks.

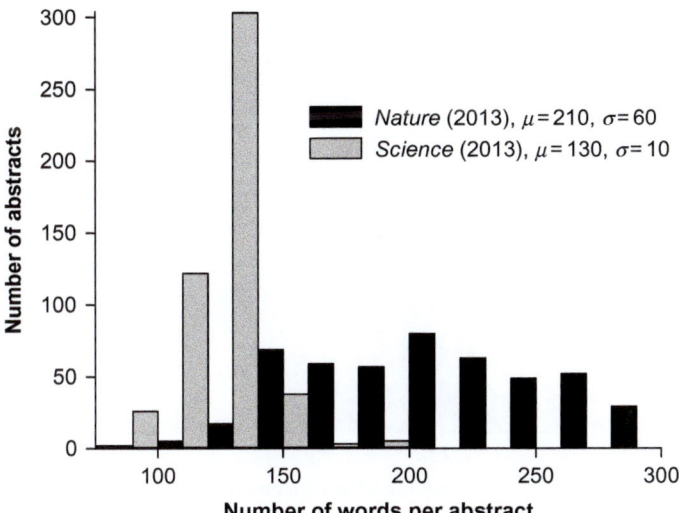

FIGURE 6.2 The 500 most cited articles in *Nature* and the 500 most cited articles in *Science* averaged 210 words and 130 words, respectively.

Abstracts in *Science* and *Nature* adhere to this format, with only five sentences (130 words) and eight sentences (220 words), respectively (Figure 6.2). The sentences average more than 22 words (Figure 6.3), which makes them longer than the target of 15 that Scott (1989) advocates. They do not fear the personal pronoun *we* or the possessive pronoun *our*: in a sample of the top 500 papers in *Nature* in 2013, *we* appears 966 times and *our* appears 238 times. Over 90% have the combination *here we*. This construction is less common in *Science*, but of its 500 most cited articles in 2013, *we* appears 681 times and *our* appears 118 times in the abstracts.

Pinker (2014) admonishes papers that confuse the research subject and the process of research. He provides this example:

> *In recent years, an increasing number of psychologists and linguists have turned their attention to the problem of child language acquisition. In this article, recent research on this subject will be reviewed.*

It is irrelevant that psychologists have now decided to investigate how children learn to speak. Saying that they will review the research is also pointless. The subject is "how children learn to speak without formalized lessons" (Pinker, 2014).

The abstracts in *Nature* and *Science* risk substituting the research as an object for what the researchers have done. The personal pronoun *we* forces the active voice, which is good, but the agent becomes the researchers and the patient

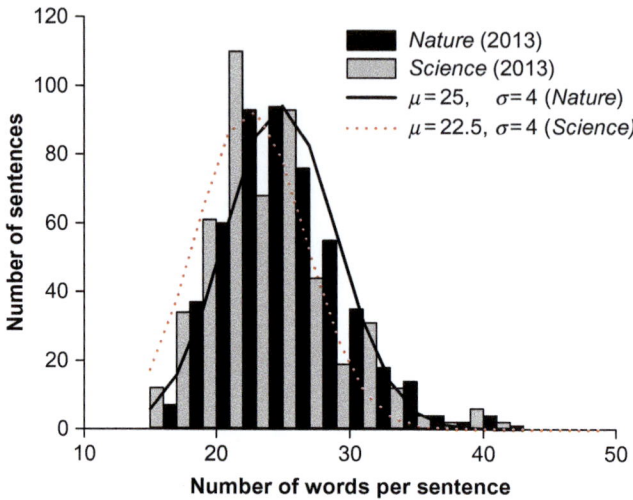

FIGURE 6.3 The 500 most cited articles in *Nature* and the 500 most cited articles in *Science* averaged 25 words per sentence and 22 words per sentence, respectively.

is what they did. Pinker (2014) asserts that readers are more interested in the results and are less interested in how researchers spend their time. *We observed, we analyzed, we identified, we investigated, we applied,* and *we found,* are the activities leading to the subject of the paper. *We present, we report, we describe,* and *we propose* are self-evident. *Our data, our findings, our classification, our method,* and *our technique,* if not contrasted immediately against other data, are hedges. These statements are feeble—be direct and assertive instead. State what the data teaches rather than qualifying it and making you or your data the agent. Keep your abstract on topic. If it contains any of the words in Table 6.1, rewrite the sentences.

Science abstracts are five sentences:

First and second sentences The first couple of sentences are among the hardest to write in the entire manuscript. With two sentences (or one long sentence of more than 25 words) *Science* authors introduce their subject and why it is important. In *Science* volume 342 issue 6165, some authors highlight challenges or controversies; others reword the title.

Title	Abstract
Photon-mediated interaction between distant artificial atoms van Loo et al. (2013, p. 1494)	Photon-mediated interactions between atoms are of fundamental importance in quantum optics, quantum simulations and quantum information processing.
Relaxation mechanism of the hydrated electron Elkins et al. (2013, p. 1496)	The relaxation dynamics of the photoexcited hydrated electron have been subject to conflicting interpretations.

Dynamical resonances accessible only by reagent vibrational excitation in the $F + HD \rightarrow HF + D$ reaction Wang et al. (2013, p. 1499)	Experimental limitations in vibrational excitation efficiency have previously hindered investigation of how vibrational energy might mediate the role of dynamical resonances in bimolecular reactions.
Brood parasitism and the evolution of cooperative breeding in birds Feeney et al. (2013, p. 1506)	The global distribution of cooperatively breeding birds is highly uneven, with hotspots in Australasia and sub-Saharan Africa.
Revealing nature's cellulase diversity: The digestion mechanism of *Caldicellulosiruptor bescii* CelA. Brunecky et al. (2013, p. 1513)	Most fungi and bacteria degrade plant cell walls by secreting free, complementary enzymes that hydrolyse cellulose: however, some bacteria use larger enzymatic assemblies called cellulosomes, which recruit complementary enzymes to protein scaffolds.

Second and third sentences After the synopsis of why the subject is important, highlight the method. *Nature* and *Science* introduce this section with metadiscourse: *Here we report, find, present, describe, propose, etc.* Sentences following this statement have relevant information. Avoid stating laundry lists of instruments and techniques. The results of the analysis are important; the instrument is secondary.

Sentence	**Comment**
The electrodes were characterized by CVA, SEM, X-ray microanalysis, XRD, XPS and ICP-AES. Kusnetsov et al. (2014, p. 829)	Abstracts focus on data that substantiate the work—leave equipment lists for the methods section.
The specific surface areas of these monoliths were determined by nitrogen adsorption, and their chemical structure was characterized using Raman spectroscopy. Goodman et al. (2013, p. 291)	Goodman et al. (2013) should state the surface areas. Experimental techniques are secondary.

Fourth and fifth sentences In the last two sentences, discuss the work quantitatively.

Sentence	**Comment**
XRD analysis of the as-deposited and heated coatings revealed that their crystallinity improved as heat treatment temperature increased. García et al. (1998, p. 69)	Higher temperature improved the as-deposited and heated coatings' crystallinity (XRD). (There is no need to say the analysis revealed an improvement.)

Sentence	Comment
This study greatly expands the genomic representation of the tree of life and provides a systematic step towards a better understanding of biological evolution on our planet. Rinke et al. (2013, p. 431)	*Greatly* is vague (a booster). How much does it expand? What is a systematic step, and how much better is our understanding?
Observation of a Hofstadter spectrum in bilayer graphene means that it is possible to investigate emergent behavior within a fractal energy landscape in a system with tunable internal degrees of freedom. Dean et al. (2013, p. 598)	Dean et al. (2013) could state this idea more clearly by deionizing the noun: *observing a Hofstadter spectrum…*
Related proteins with PrLDs should therefore be considered candidates for initiating and perhaps propagating proteinopathies of muscle, brain, motor neuron and bone. Kim et al. (2013, p. 467)	Related proteins with PrLDs are candidates to initiate and propagate…(*Candidates* already implies doubt. There is no need to hedge further.)
Our results identify the human Mediator complex as the transducer of activating ncRNAs and highlight the importance of Mediator and activating ncRNA association in human disease. Lai et al. (2013, p. 497)	Lai et al. (2013) are hedging; they blame it on their results. *The human Mediator complex is the transducer of…*

6.3 GRAPHICAL ABSTRACT

The graphical abstract and research highlights are the most effective constructs to communicate the essential details of your manuscript. Many people see them since they are included with the title and authors on the table of contents page of many journals' websites. Eye-catching graphical abstracts that are self-explanatory encourage people to spend a couple of seconds looking at them. This might incite them to read the article. The quality of many graphical abstracts is poor because people neglect to spend time making them. They are the *Oh shit* when you finished uploading the last document to the journal's website. Although it might be the last thing we do, it is the first thing everyone sees. We might spend between 2 h and 4 h on the graphical abstract (and less even for the highlights), whereas preparing the manuscript takes 100 h.

As you prepare your manuscript, think about how to compose the graphical abstract to entice people to read your paper. Identify two or three images that represent your work best. Illustrations and text taken directly from your manuscript can become illegible if you copy them directly for the graphical abstract. The geometry of graphs in a paper ranges from 1:1 to 3:2. The window in graphical abstracts is 5:2. Specific dimensions may vary according to the journal. The area in *Acta Biomaterialia*, for example, is 500 px wide by 200 px high. Choosing a 1:1 illustration for the graphical abstract will result in a 200 px × 200 px image. By using the full 500 px × 200 px format, you have 6000 more pixels to convince readers to read your paper. *Chemical Engineering Science* specifies that the graphical abstract be at least 1328 px × 531 px. Further, *Chemical Engineering Science* states that it should be legible at a screen resolution of 96 dots per inch (dpi) and 13 cm × 5 cm. The graphical abstract must be proportional to these dimensions; otherwise the graphic designer has to squash the vertical axis to fit it in the window. Characters and symbols disappear when they are squashed too much. Because you see it well on your screen doesn't mean that it will look good on the journal's website.

If journals maintained the illustrations as vector images, fitting the image into a smaller window wouldn't be traumatic because readers can zoom in and the quality remains the same. However, journals convert the graphical abstracts to raster images, so quality suffers. Zooming in doesn't help. What was clear in the article becomes blurry on the website. Convert your vector images to raster images to check them.

In Chapter 4, we recommend that text in graphs be about the same size as text in the caption or body of the manuscript. To ensure that your text is legible, choose a font size such that the text occupies 5% to 10% of the vertical axis. *Acta Biomaterialia* suggests a font size of 10 pt to 16 pt, but this guideline ignores how the graphic designer will scale the image to fit in the window. Figure 4.9 illustrates the number 9 with a 7 px vertical resolution. The quality is poor, but the number is legible.

The conductive hydrogel image of Cheong et al. (2014) that *Acta Biomaterialia* published is in 1:1.3 format (Figure 6.4(a)). The caption font is 2 pt smaller than the text font, and the font in the figure is another 1 pt smaller. All text is legible in the paper.

The authors submitted this figure as the graphical abstract. The text height is less than 2% of the height of the image. Converting the figure to 200 px (vertically) means that the pixel resolution of the uppercase letters becomes $0.02 \times 200 = 4$ px (for the upper case letters and numbers). The resolution of the lowercase letters is 2 px, and they are illegible (Figure 6.4(b)).

Dorj et al. (2014) did a good job combining three photos and one graph to form their graphical abstract. They related a photo to the graph with an arrow and added text to highlight the relationship between different elements. Unfortunately, almost all of the text is too small because they stacked their images vertically instead of spreading them out horizontally. Some numbers and uppercase letters are 1.7% of the vertical height (Figure 6.5(a)). Even at the highest magnification, the text is illegible. In Figure 6.5(b), we converted the stacked, square geometry to fit *Acta Biomaterialia*'s guidelines and added text that is 5% of the vertical height. We also modified the plot to fit better with

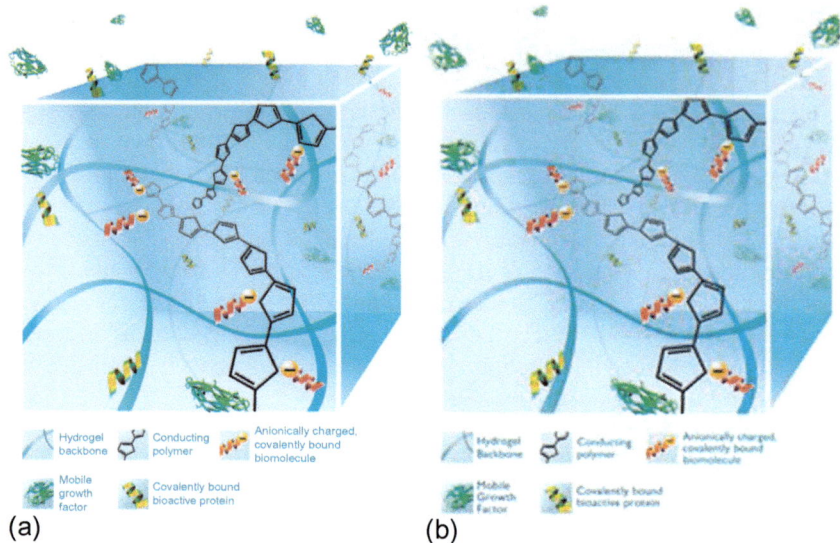

(a)   (b)

FIGURE 6.4   Conductive hydrogels with tailored bioactivity. (From Cheong et al. (2014).) (a) This image is from the journal article (a pdf) on Acta Biomaterialia's website. The vertical resolution of the capital letter C in the label is greater than 60 pixels. (b) This image is the GA on Acta Biomaterialia's website. The vertical resolution of the capital letter C in the label is 5 pixels. The image quality is poor and many characters are illegible.

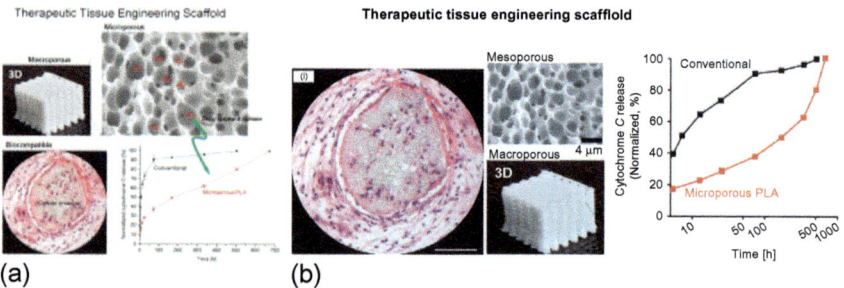

(a)   (b)

FIGURE 6.5   Biopolymer scaffolds for drug loading and delivery. (From Dorj et al. (2014).) (a) The GA of Acta Biomaterialia is 500 pixels by 200 pixels. (b) We modified the scatter plot geometry to place the images on a horizontal plane. We also rewrote all text so the numbers and capital letters are 5% of the GA height.

the three photos that Dorj et al. (2014) selected, and adopted a $\log_{10}$ scale for the *x*-axis to spread out the data points. To compare the figures, we chose the height of the original graphical abstract as the height of our version.

The graphical abstract of Valadés-Pelayo et al. (2015) is a good example of how to use space and combine images with text to develop a story: it starts on the left side, and the arrows combine images and bifurcate (Figure 6.6).

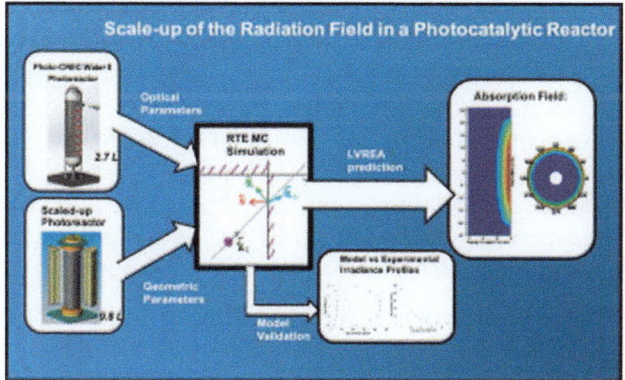

FIGURE 6.6 Flowchart graphical abstract: scale-up of the radiation field in a photocatalytic reactor. (From Valadés-Pelayo et al. (2015).)

However, *Chemical Engineering Science* requested images proportional to 5:2. This graphical abstract is in 3.3:2 format; their image was clear, but when the journal converted it to a raster image and shrunk it, the text and the graphs became illegible. Black text on a white background is better than white text on a blue background, and the authors could enlarge the images without framing them with black lines (with different weights).

Tips

- Take the time to make the graphical abstract fit into the space.
- Modify the font so that uppercase letters are at least 5% of the height of the graphical abstract.
- Reorient x-y plots so that the y-axis is longer than the x-axis.
- Minimize dark images.
- Add color (Chapter 7).
- Label images with text so that they are self-explanatory.
- Don't repeat the title in the text.
- Link related elements with arrows.

6.4 RESEARCH HIGHLIGHTS

The abstract is the principal source of information for the highlights. In fact, highlights are redundant when the abstract is well written, but they appear in the contents lists, in online search engines, and at times together with the graphical abstract. The highlights can be complete sentences that are self-sustaining and include an active verb or past participle. Elsevier B.V. (2014) describes them as short sentences that convey the essence of the research—conclusions or results:

"Get to the point with research highlights. These are bulleted lists of key findings of an article to help readers see whether the research is of interest to them." Usually there are five highlights, which correspond to the five sentences in the abstracts of *Science* (but with half the number of words).

Repeating the title of the paper as part of the highlights is redundant; however, if search engines ignore titles, this might be necessary. Here is an example of sequential highlights that reproduce the paper's title: *The function of the respiratory supercomplexes: The plasticity model* (Acin-Perez and Enriquez, 2014). Each subsequent bullet point is related to the previous, like in a paragraph.

- The plasticity model explains the organization of respiratory complexes (RCs).
- It proposes that RCs can be free or super-assembled with other complexes.
- The dynamic supra-organization of the RCs is a short-term regulation of metabolism.
- Fuel utilization is optimized by mitochondrial supercomplex dynamics.

Like an abstract, highlights should communicate substantive information. Assertive statements are best. The following highlights are too general:

- Investigation of n-6 fatty acids in the prevention of cardiovascular disease.
- Heat loss and thermal balancing explored along with performance and emission.
- A model for analysis of thermogravimetric experiments.
- Reaction kinetics, particle morphology and mass transfer predict performance.
- To investigate the particle size distribution, a population balance model was developed.
- Sample skewness and excess kurtosis were used to classify the fluidized bed behavior.

Minimize qualitative statements and focus the highlights on factual and quantitative results.

Exhaust heat loss decreases with the increase of biodiesel in the blend.	Increasing the biodiesel content in the blend up to 20% decreases heat loss by 10%.

Tips

- Don't repeat the title.
- Use abbreviations and acronyms that anyone can deduce from the title. Otherwise spell out any acronym you didn't learn in high school.
- Be assertive and use active verbs.
- Apply the same rules for writing a title and an abstract.

6.5 INTRODUCTION

While preparing your introduction, bear in mind what Joseph Pulitzer said about writing:

Put it before them briefly so they will read it, clearly so they will appreciate it, picturesquely so they will remember it and, above all, accurately so they will be guided by its light.

Pinker (2014) endorses Pulitzer and further advocates writing in the classic style in which the writer engages the reader in a conversation. He considers that writing establishes a relationship between the author and the reader. The author recognizes that the reader is competent, and must write clearly and simply so that the reader recognizes the truth.

When the introduction is bad, by extension, reviewers and the intended readership are justified in assuming that the rest of the paper is equally bad. Spell correctly, check the grammar, and discard overused banal statements such as: "The concentration of greenhouse gases is increasing." Don't make broad generalizations or embellish controversies that validate your point of view.

The abstract introduces the subject, asserts facts, and highlights significant results of the work. You establish a relationship, your credibility, with the reader in the introduction: you support what you assert in the title and abstract with coherent arguments and facts. State clearly the context of the work and why the work is important, and substantiate any assertions with authoritative references. Discuss some of the details of the work, but also set the scope of the paper. Introduce readers to the historical context, the economic incentive, and the scientific interest. The introduction complements the first couple of sentences of the abstract, and in many journals reproduces them closely. In their abstract, Green and Alemseged (2012) state that the form of primates' shoulders relates to how much they climbed. They repeat this concept in the first two sentences of the introduction, and add seven references.

Abstract	**Introduction**
Scapular morphology is predictive of locomotor adaptations among primates, but this skeletal element is scarce in the hominin fossil record. Green and Alemseged (2012, p. 514)	Scapular morphology corresponds closely with locomotor habits, often irrespective of phylogeny (1-7). However, our understanding of this important element in hominin evolution is limited by the paucity of scapular fossil remains.

Reviewers appreciate a thorough survey of the literature that highlights important contributions of other researchers in the field.

Literature Review

Reviewing literature is a continuous activity. Mention the major contributions, any controversy, and what is left to do. A critical review requires a couple of sentences for each reference to describe the salient features of the previous work and its limitations. When several articles touch on the same subject in a similar manner, they can be referenced simultaneously: $[\text{ref}_x - \text{ref}_{x+y}]$.

The literature review substantiates why you undertook the research. Any statement you begin with *It is suggested that...* demonstrates apathy toward identifying who said what first or that you don't grasp the literature. It is insufficient to cite the latest research papers or refer to papers that support your analysis. But you also have to be careful with older authoritative papers and books, because they may contradict the latest research.

When you review the literature thoroughly and cite previous work adequately, you show that you master your subject, and your paper earns credibility. Citing recent references demonstrates that your work is pertinent and original. References are a source of inspiration while you do your research and write articles. They are also useful to readers who wish to contextualize your work. They introduce readers to related subjects.

When you reference procedures and equipment detailed in other papers, you save space.

Reviewers are selected from among the people who work in the field, from the people you reference, and those who you have not referenced. A reviewer whose work you have overlooked might be overly critical of your work. On the other hand, if reviewers recognize that you appropriately represent their work, they have an interest in the journal publishing the article. Cite work that contradicts your own work or that you demonstrate has errors.

Search engines indicate immediately when your paper is referenced, and so by citing more people, you have a larger audience. We stated above that people first read the abstract and then the conclusions. In fact, many will check the references first to see if you cite their work. So, more references are better (up to a point).

Citing 10 articles or fewer is presumptuous. As we discuss in Chapter 9, citing a meager number of references may constitute plagiarism. Besides, citing more papers increases the likelihood that people in your field will read your paper. Review articles cite hundreds of papers, but scientific papers cite fewer papers. As a benchmark, we compiled 3200 articles and letters that *Nature* published between 2011 and 2014. They averaged 36 references per paper. One of the references was from 1687, 4 were from the 1700s, and 57 were from the 1800s.

The age of the reference is the number of years between when *Nature* published the article and when the reference was published. The number of references, $n(t)$, decreases exponentially and follows a Weibull distribution (Figure 6.7):

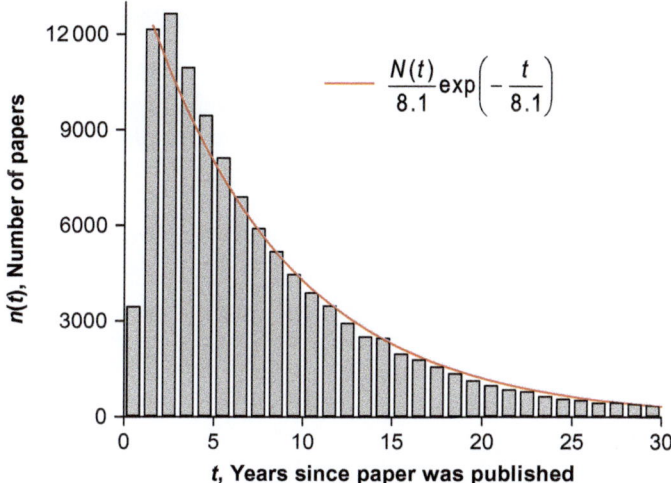

FIGURE 6.7 Articles and letters published in *Nature* between 2011 and 2104 referenced 112 839 papers. The frequency distribution of the references cited in the article versus the age of the article follows a Weibull distribution with a modulus (shape parameter of slope) of 1 and a scale parameter of 8.1 ($R^2 > 0.99$).

$$n(t) = N\frac{m}{\theta}\left(\frac{t}{\theta}\right)^m \exp\left[-\left(\frac{t}{\theta}\right)^m\right],$$

where N is the total number of citations from the time the oldest reference was published, m is a shape parameter (also known as the modulus or the slope) and θ is a scale parameter: 63% of the references are younger for $t < \theta$.

Over 3200 references appeared the same year that *Nature* published its papers; we assigned $t = 0.5$ years to these references. Twelve thousand references appeared 1 year earlier ($t = 1.5$ years) and an equal number appeared 2 years earlier ($t = 2.5$ years). t_{50} is 5.5 years, which means that 50% of the references are less than 5.5 years old. Forty percent are less than 3.5 years old, and 17% are more than 10.5 years old. The average age (assuming a normal distribution) is 8.6 years.

$$n(t) = 112\,839\left[1 - \exp\left(-\frac{t}{8.1}\right)\right]$$

The scaling parameter is 8.1, which means that 63% of the referenced articles were published within the previous 8.1 years.

You must cite your own work, but if over 20% of the cited papers are from your research group, people (and reviewers) may question your motivation. Referencing high-impact journals and citing review articles is important to reach a broad audience. However, citing review articles should not substitute for offering your own critical review of the state of the art. Patents are a rich source of information and innovation. However, the scientific literature ignores them.

Use bibliographical software (BibTeX, EndNote®, etc.) to manage your references so you reduce omissions and typos. These reference software platforms are excellent tools to automatically format your citations for different journals and books.

Stating the Novelty and the Objectives of Your Work

A thorough literature review helps to substantiate the originality of your paper, but remember to state the originality explicitly in the first part of the introduction. The literature review outlines what we know about the subject. After that, state how the work contributes to advancing knowledge. Pinker (2014) recommends minimizing signposting—telling the reader what is coming next—which is a form of metadiscourse.

The following example repeated the last two sentences of the abstract as the last two sentences of the introduction. The authors use signposting unnecessarily. Their purpose—to present a derivation—is rather modest.

Abstract—last sentences	**Introduction—last sentences**
The purpose of this report is to present the derivation, assumptions, and applications of the $2^{-2\Delta\Delta C_T}$ method. In addition, we present the derivation and applications of two variations of the $2^{-2\Delta\Delta C_T}$ method that may be useful in the analysis of real-time, quantitative PCR data. Livak and Schmittgen (2001, p. 402)	The purpose of this report is to present the derivation of the $2^{-2\Delta\Delta C_T}$ method, assumptions involved in using the method, and applications of this method for the general literature. In addition, we present the derivation and application of two variations of the $2^{-2\Delta\Delta C_T}$ method that may be useful in the analysis of real-time quantitative PCR data.

6.6   METHODS

The methods section has the same obligation as a patent: it must provide enough information such that a person skilled in the art can reproduce the data. It necessarily describes the experimental equipment and specialized analytical instruments, lists important materials and reagents, and summarizes the procedure or theoretical calculations.

Include professional schematic diagrams, and describe major pieces of equipment individually. Describe the experimental sequence chronologically and the sampling procedure, if any. List the conditions of the experimental design, the reagents, and the materials in tables. State the invariant factors in the body or in the captions to keep tables manageable. Indicate if you further purified the reagents or synthesized them as part of the study. Reference the synthesis steps, where appropriate, rather than repeating them, but state any differences. Provide the brand and the model of the analytical instruments as well as the conditions.

Experimental sections abuse the passive voice. It is easier to write in the passive voice but it takes more text to say the same thing than does the active voice. Here are expressions that you should discard from your repertoire and find a way to say with the active voice:

— experiments (measurements) were carried out (done at, performed)
— instruments were used to
— was prepared by
— was utilized to
— was investigated (analyzed) by
— was quantified by (in order to quantify)
— the instrument was placed
— made measurement
— performed an experiment
— was maintained
— was used for the determination of (to acquire information)
— this approach allows for accurate interpretation
— The experimental apparatus is illustrated in Figure #

Make the instrument the agent. This goes for pumping, measuring, heating, drying, etc.

The fluid was pumped through the coils using a positive displacement pump connected to a variable speed motor. The speed of the motor was adjusted to obtain flow rates of 0.10, 0.15 and 0.20 kg · s^{-1}. Prabhanjan et al. (2004, p. 36)	A positive displacement pump metered the fluid through the coils at 0.10, 0.15 and 0.20 kg · s^{-1}.

Was pumped is passive. The implicit agent is *we*, but you can improve this sentence if you make the pump the agent. However, saying that a pump pumps the fluid sounds redundant. Change the verb *pump* to one of the alternatives *feed*, *meter*, *send*, *inject*, or *force*. The variable speed motor is insignificant information. It's better to provide the quantitative information in the same sentence that you describe the action, so mention the rates in the first sentence. The corrected version is better because it gives the same information in one sentence instead of two.

When you describe your analytical instruments, state the model and brand name. Also, make the instrument the agent in the sentence, but don't say that it used the analytical technique.

Original	**Corrected**
The transmission electron microscopy (TEM) was performed with a Jeol [sic] electron microscope (model JEM 100CX II). Wang et al. (2005, p. 112)	A JEOL electron microscope (model JEM 100CX II) imaged the samples (TEM).

X-ray photo-electron spectroscopy (XPS) was utilized to monitor the presence of residual elemental impurities in the final products and to study the chemical state of the GO sheets. The data were gathered using a hemispherical analyzer supplied by an Al Kα X-ray source ($hv = 1486.6\,\text{eV}$) operating at a vacuum higher than 10^{-7} Pa.	A hemispherical X-ray photo-electron spectrophotometer (XPS) with an Al Kα X-ray source ($hv = 1486.6\,\text{eV}$) operating at a vacuum higher than 10^{-7} Pa monitored residual elemental impurities in the final products.

Akhavan et al. (2014, p. 20442)

The implicit agent is *we*, but it's better if you make the instrument the agent. Don't say that it used the technique. Obviously it did. Instead state what it did with your samples.

6.7 RESULTS AND DISCUSSION

Some journals combine the results and discussion sections; others have a distinct section for each, where the results section presents the experimental data and the discussion section interprets it. This section substantiates what you say in the abstract and conclusions. It constitutes the bulk of a paper—the raison d'être. Start by describing the philosophy of the experiments or model. State the implications of the data. Include literature references to corroborate your results. If the results are inconsistent with the literature, highlight the differences. Explain the results, graphs, and tables in context.

Stating trends in figure or table data is the first step to describe the data, whether or not it is obvious. This introductory sentence draws the reader's attention to specific phenomena. Then you relate the trend to a physical phenomenon or the main thesis of your paper. Also,

- quantify the experimental errors,
- put into context the experimental procedure and results,
- specify the philosophy of the experimental sequence,
- acknowledge limitations of the study,
- highlight the originality of the data or model,
- discuss the relationship of the data with existing literature,
- describe how the work applies to other fields,
- interpret the results in the context of a mathematical or physical model,
- identify practical implications of the work, and
- recommend new avenues of research.

Describe, highlight, or discuss every table and figure in the manuscript. If you don't mention a figure or table in the text, remove it (or discuss it). Don't begin sentences with *Table # summarizes* or *Figure # shows*, which is metadiscourse. Discuss the data. Mentioning in the text some numerical data from a table is insufficient.

Deficient	**Acceptable**
Figure 1A shows the XRD pattern of commercial HA powder (MetcoB) sintered at 1250 °C. Minimal line broadening can be seen, which indicates a well-crystallized material. Wolke et al. (1994, p. 1479)	The commercial HA powder (MetcoB) sintered at 1250 °C. It is well crystallized: the line broadening in the XRD pattern is minimal (Figure 1A).
Figure 1 shows that even when using the original one-hit method with threshold parameter $T = 13$, there is generally no greater than a 4% chance of missing an HSP with score > 38 bits. Altschul et al. (1997, p. 3391)	The chance of missing an HSP with a > 38 bit score is 4% or less even with the original one-hit method with threshold parameter $T = 13$ (Figure 1).
In Table 1 we can see that the results obtained for the variance are much larger than those of the mean, suggesting that the hypothesis that all articles have the same expected impact is wrong as we might have guessed. Vieira and Gomes (2010, p. 3)	The variance is larger than the mean, which confirms our expectation that all the articles have a different impact (Table 1).
High-resolution micrographs of typical needles show {002} lattice images of the graphite structure along the needle axes (Fig. 1). Iijima (1991, p. 56)	(The original is good; we don't need to correct it.)

WoS recorded 47 000 citations to Becke's (1993) work, an article on density functional thermochemistry, up to January 2015. It was the most cited paper, not because it was well written but because

- it proposed a new theory with a superior precision,
- the field of density functional theory was becoming popular,
- the theory had wide applications to both physics and chemistry, and
- the paper contained an extensive data set.

Your work will probably never be cited as often as this one. However, write well to increase the probability of it being cited more often.

6.8 CONCLUSIONS

People first read the abstract and then read the conclusions. So repeating in the conclusions what is already in the abstract is redundant. Avoid restating the problem and the context of the work, but highlight the most significant findings. Address the implications of the research—extrapolate to other systems, identify potential applications, mention future experiments or studies, and report general and specific limitations of the work.

Many top journals—*Science, Nature, Proceedings of the National Academy of Science of the United States of America*, for example—don't have a conclusions section. The *New England Journal of Medicine* has an extended abstract with "Background," "Methods," "Results," and "Conclusions" subheadings. On the other hand, *Physical Review B* qualifies the conclusions as "Conclusions and Outlook," and *Applied Mathematical Modelling* has "Concluding Remarks."

Here are several good conclusions:

1. We have revealed in this work the important and potentially powerful role of exact exchange in density functional thermochemistry. The semiempirical combination of exact exchange, the LSDA, and gradient correction terms in Eq. (2) performs significantly better than the gradient-corrected exchange-correlation functional of paper II (Ref. 5), and approaches an average precision of order 2 kcal/mol (or 0.1 eV, $10 \, \mathrm{kJ \, mol^{-1}}$). (Becke, 1993)
2. It is surprising that SHELX has been able to maintain a dominant position in small-molecule structure determination for the last 30 years, even though computers have changed out of all recognition in this time. (Sheldrick, 2008)
3. The main results, shown in Figs. 3 and 5 point out some intriguing opportunities offered by plasmons in graphene for the field of nanophotonics and metamaterials in infrared (i.e., for $\omega < \omega_{Oph}$). For example, we can see in those figures that high field localization and enhancement $\lambda_{air}/\lambda_p \sim 200$ [see Fig. 3(b)] are possible (resulting in $\lambda_p < 50 \, \mathrm{nm}$, while plasmons of this kind could have propagation loss-lengths as long as $\sim 10\lambda_p$ [see Fig. 5(a)]; … Another interesting feature … (Jablan et al., 2009)
4. Together with fatty acid methyl esters, we detected products resulting from cracking reactions, which potentially can improve the cold flow properties. Furthermore, because of the high temperatures, the process is suitable to treat vegetables oils, waste cooking oils and even yellow and brown grease. The economic incentive of gas phase transesterification goes beyond high conversion (since conversion is not limited as in the liquid phase process), purity, low cost feedstocks but includes very high weight hourly space velocities. (Boffito et al., 2013)
5. Our analysis does have some general limitations, such as the data quality of health outcomes relevant to all global studies—i.e., mortality and burden of disease and specific limitations for a CRA for alcohol (Murray and Frenk[53] provide a general discussion of the limitations). Furthermore, we refer only to ongoing discussions on DALY assumptions[17] including the derivation of the DALY weights,[54] and on data quality for global mortality.[19]

   Our approach also has some specific limitations. (Rehm et al., 2009)
6. The incubation period and case fatality rate among patients with EVD in Sierra Leone are similar to those observed elsewhere in the 2014 outbreak and in previous outbreaks. Although bleeding was an infrequent finding, diarrhea and other gastrointestinal manifestations were common. (Schieffelin et al., 2014)

Becke (1993) identifies what was important in his work. Jablan et al. (2009) *point out intriguing opportunities*, they identify fields of applications, and identify interesting features. These are all excellent characteristics of a well-written conclusion/outlook. Boffito et al. (2013) mention advantages of their technology. Rehm et al. (2009) report the limitations of their work.

Avoid general statements such as *more experiments are required*. Rather, indicate what the focus of additional work should be. Boffito et al. (2013) presume that the process is suitable for waste cooking oils. They hadn't done the experiments to prove this statement, so we understand that more experiments are required. Another example is: *A comprehensive mathematical model has been developed*. It would be obvious from the abstract and the body of the paper that the authors developed a mathematical model. Rather than repeat yourself, reiterate how important the model is or how it applies to other fields of research, etc.

If you have any of the following phrases, rewrite them:

— …showed good agreement
— …was investigated
— …has received considerable attention
— …An experimental…was conducted.
— It is shown that…
— It is proven that…
— It is demonstrated that…
— The model proposed successfully…
— From the work presented here, one can conclude that…
— In this work…were compared

We extracted the following list of verbs (sequentially) from the conclusions of a paper that we reviewed. We suggested that the authors identify the patient and agent for every sentence. As written, the implied agent for most of the sentences is the author. Adding *we* to make them active improves them, but only marginally.

— are used
— is found
— is proposed
— is seen
— is predicted
— are identified
— is constructed

6.9 ACKNOWLEDGMENTS

Acknowledgments list individuals that contributed substantially to the manuscript, institutions that supported the work financially (or in-kind contributions), and potential conflicts of interest. Authors participate more in the article than contributors, and are legally accountable for the work. Universities require students and collaborators to substantiate their contribution to an article as a percentage of the overall effort. In some cases, students must allege that they contributed more than 75% of the work of the paper. When a patent is commercialized, inventors receive royalties proportional to how much they contributed.

When many researchers work on multi-institutional studies, assigning how much each contributes is arbitrary (and even with more than five researchers). That is less important than deciding who should be considered as a coauthor. The International Committee of Medical Journal Editors (2014) recognizes four criteria for authorship. An author must have:

1. conceived and designed the study or parts of it, or collected, analyzed, or interpreted data;
2. drafted the article or revised important parts of it;
3. approved the article's final version; and
4. agreed to be accountable for the results.

Some journals want to know who did what. The Royal Society of Chemistry (2014) requests a list of coauthor contributions for papers with over ten authors. The American Association for the Advancement of Science (2014) asks coauthors how much they participated (in increments of 20%) for criteria similar to those of the International Committee of Medical Journal Editors (2014). Moreover, senior authors from each research group must confirm that they checked the data their laboratory generated. In fact, all authors should be aware of the contributions of all the work undertaken by others. This criterion might be difficult to meet in multi-institutional research projects such as that which resulted in the article on intensive treatment of diabetes published by the *New England Journal of Medicine* with 681 authors (Shamoon et al., 1993). (Several authors in this study are listed twice.)

The International Committee of Medical Journal Editors (2014) recommends that editors oblige authors to identify their contributions.

- V.C. conceived the project, performed the survey, and processed the data. (Calcagno et al., 2012, p. 1069)
- V.C. analyzed the data with inputs from E.D., K.G., D.R., and C.d.M. and wrote the manuscript with all authors. (Calcagno et al., 2012, p. 1069)
- S.T.S. conducted the impact simulations, M.Ć. calculated the tidal evolution, and both wrote the paper. (Ćuk and Stewart, 2012, p. 1052)
- Conceived and designed the experiments: JPAI. Performed the experiments: KWB. Analyzed the data: JPAI KWB RK. Contributed reagents/materials/analysis tools: KWB RK. Wrote the paper: JPAI. (Ioannidis et al., 2014)

Listing contributions in the supplementary materials shortens the acknowledgments.

- Accession numbers and author contributions are listed in the supplementary materials. (Sprunck et al., 2012, p. 1097)

Individuals who collected data but didn't write anything aren't authors: they are contributors. Merely funding the research, administrating, or proofreading is insufficient to qualify you as a coauthor. Overall thanks are appropriate in many cases, but adding individual names together with the overall thanks is better:

- Finally, we thank all of the people who have been involved with various CLUSTAL programs over the years, namely Paul Sharp, Rainer Fuchs, and Alan Bleasby. (Thompson et al., 1994)

It's obvious that the infrastructure of your department contributes to all your work. Furthermore, you've already recognized your department in the author affiliations.

- We thank T. Coupet, D. Meyer, and E. Pryce for technical assistance; R. Coleman, Y. Wang, C. Arrowsmith, and members of our department for sharing their equipment and materials; and A. Le, E. O'Shea, and members of our department for comments on the manuscript. (Keller et al., 2012, p. 1072)

Acknowledging *members of our department* is redundant.

Acknowledge peers who contributed significant advice, criticism, expertise, or assistance. Remember to ask for permission before acknowledging someone.

- We thank L. R. Kroos and L. R. Snyder for helpful discussions and comments on the manuscript and K. Lewis and J. Landgraf for technical assistance. (Somvanshi et al., 2012, p. 93)
- We wish to specifically thank Jeroen Coppieters who suggested using a series of weight matrices and Steven Henikoff for advice on using the BLOSUM matrices. (Thompson et al., 1994)

Fujii and Zwart (2011) thank anonymous referees for their criticism. This may seem pointless because nobody but the referees (and the editor) know who they are. However, it's gratifying for referees to know that the authors recognized their contribution.

- We thank B. Brandl, A. Brown, A. Gürkan, K. Nitadori, and H. Sana for discussions and the anonymous referees for their constructive criticism. (Fujii and Zwart, 2011, p. 1383)

If you thank someone from one of the institutions participating in the study, you need not specify the institution. Otherwise, specify the affiliations to clarify potential conflicts of interest.

- We acknowledge collaboration with H. Nakanishi (Toyota Motor Corporation). (Suntivich et al., 2011, p. 1385)

The authors of this work are from MIT and the University of Texas, and they mentioned Nakanishi's affiliation to Toyota.

Acknowledge resources, facilities, and software, including advanced analytical instruments.

- The research made use of the Shared Experimental Facilities supported by the MRSEC Program of the National Science Foundation under award DMR 08-019762. (Suntivich et al., 2011, p. 1385)
- The impact simulations were run on the Odyssey cluster supported by the Harvard Faculty of Arts and Sciences Research Computing Group. (Ćuk and Stewart, 2012, p. 1052)

When you acknowledge financial support such as grants, scholarships, and fellowships, state to which author the funds went. Some articles acknowledge financial support before mentioning other contributors.

- This study was supported by Michigan State University Research for Excellence Fund Center for Microbial Pathogenesis, AgBioResearch, and startup funds (to T.A.C.) and by NIH (grant R01 GM086258 to J.C.). (Somvanshi et al., 2012, p. 93)
- This work was supported by a Natural Environment Research Fellowship award (to B.R.), the Biotechnology and Biological Sciences Research Council (grant no. BBC 5127021), the European Research Council, and the Royal Society. (Raymond et al., 2012, p. 88)

Conflicts of interest include employment, consultancies, stock ownership or options, honoraria, patents, and paid expert testimony (International Committee of Medical Journal Editors, 2014). Don't mention personal relationships or rivalries, academic competition, and intellectual beliefs (International Committee of Medical Journal Editors, 2014). Conflicts of interest statements clarify your position: you declare your affiliations so that readers know whether you have an interest in the success of whatever you're describing.

- A patent related to this invention has been filed. (Suntivich et al., 2011, p. 1385)
- S.S. and T.D. are inventors on a patent (WO 2007/092992) related to the *EC1* promoter sequences. (Sprunck et al., 2012, p. 1097)

Don't declare that you have no conflict of interest (financial or otherwise) when you have none (unless the journal requires you to declare that you have none).

- The authors declare no conflicts of interest. (Raymond et al., 2012, p. 88)
- The authors declare no competing financial interests. (Kirste et al., 2012, p. 1063)

Although official statements and additional materials don't belong in the acknowledgments, some journals put them there rather than making a new section.

- Raw field data are available in the supplementary materials. (Raymond et al., 2012, p. 88)
- The simulation software is available at http://amusecode.org and the data can be found at http://initialconditions.org. (Fujii and Zwart, 2011, p. 1383)
- The opinions expressed herein are those of the authors and do not necessarily reflect the views of USAID. (Coppock et al., 2011, p. 1398)
- The content is the responsibility of the authors and does not necessarily reflect the official views of the NIGMS, NCI, or NIH. (Julien et al., 2013, p. 1483)

6.10 EXERCISES

1. *Title*: Which words should you remove from the following titles?
    (a) Studies on the quantitative and qualitative characterization of erythrocyte glutathione peroxidase (Paglia and Valentine, 1967)
    (b) Assessing the performance of prediction models: a framework for some traditional and novel measures (Steyerberg et al., 2010)
    (c) A dynamic model of customers' usage of services: Usage as an antecedent and consequence of satisfaction (Bolton and Lemon, 1999)
    (d) Optimized codon usage and chromophore mutations provide enhanced sensitivity with the green fluorescent protein (Yang et al., 1996)
    (e) Analysis of relative gene expression data using real-time quantitative PCR and the $2^{-\Delta\Delta C_T}$ method (Livak and Schmittgen, 2001)
    (f) The use of phospholipid fatty acid analysis to estimate bacterial and fungal biomass in soil (Frostegard and Baath, 1996)
    (g) Definitions for sepsis and organ failure and guidelines for the use of innovative therapies in sepsis (Bone et al., 1992)
    (h) A new application for microgels: Novel method for the synthesis of spherical particles of the Y_2O_3:Eu phosphor using a copolymer microgel of NIPAM and acrylic acid (Martinez-Rubio et al., 2000)
    (i) Corrosion protection by organic coatings: electrochemical mechanism and novel methods of investigation (Grundmeier et al., 2000)
    (j) Design, synthesis and investigation of mechanisms of action of novel protein kinase C inhibitors: perylenequinonoid pigments (Diwu et al., 1994)
2. *Abstract*: Improve the following sentences/paragraphs:
    (a) A new method is introduced for estimating the convection velocity of individual modes in turbulent shear flows that, in contrast to most previous ones, only requires spectral information in the temporal or spatial direction over which a modal decomposition is desired, while only using local derivatives in other directions. (Álamo and Jiménez, 2009)
    (b) The preparation of spherical phosphor europium-doped yttrium oxide $(Y_2O_3$:Eu) particles using a copolymer of N-isopropylacrylamide

(NIPAM) and acrylic acid (AAc) is described. (Martinez-Rubio et al., 2000)

(c) The water splitting experiments were conducted in a 7 mm ID quartz tube. (Chiron and Patience, 2012, p. 10529)

(d) The Water and Global Change (WATCH) project evaluation of the terrestrial water cycle involves using land surface models and general hydrological models to assess hydrologically important variables including evaporation, soil moisture, and runoff. (Weedon et al., 2011, p. 823)

(e) Lipid oxidation is, together with microbial growth, the main cause of spoilage of a great variety of foods, such as nuts, fish, meats, whole milk powders, sauces and oils. It causes a loss of both sensorial and nutritional quality of foods and may even lead to the formation of toxic aldehydes. Some strategies that are commonly used to limit the extent of lipid oxidation of packaged foods are direct addition of antioxidants or packaging under modified atmospheres in which oxygen presence is limited. A novel alternative to these methods is antioxidant active packaging, whose main advantage is that it can provide sustained release of antioxidants during storage. This article reviews the latest advances in antioxidant active food packaging, with special emphasis on antioxidant release systems. The various methods for incorporating antioxidant compounds in the package, the issues to be considered in packaging design, and the various methods employed to date to evaluate the antioxidant effectiveness of active antioxidant materials are reviewed. (Gómez-Estaca et al., 2014)

(f) Great ape gestural communication is known to be intentional, elaborate and flexible; yet there is controversy over the best interpretation of the system and how gestures are acquired, perhaps because most studies have been made in restricted, captive settings. (Hobaiter and Byrn, 2011)

(g) An ecological niche modelling (ENM) approach was developed to model the suitable habitat for the 0-group European hake, Merluccius merluccius L., 1758, in the Mediterranean Sea. (Druon et al., 2015)

(h) We present here a method to separate the Stern and diffuse layer in general systems into two regions that can be analyzed separately. (Zhang et al., 2013b)

3. *Graphical abstract*:

(a) What is wrong with Figure 6.8? (Lan et al., 2015)

(b) How would you improve the graphical abstract in Figure 6.9? (Picard and Burelle, 2012)

(c) What is missing in the graphical abstract in Figure 6.10?

(d) Graphical abstracts in the *Journal of Polymer Science Part A: Polymer Chemistry* are attractive. The journal adds space to the left of the *graphical abstract* to add text—essentially a highlight. The highlight for Figure 6.11 reads "New photo-induced thiol-ene crosslinked films

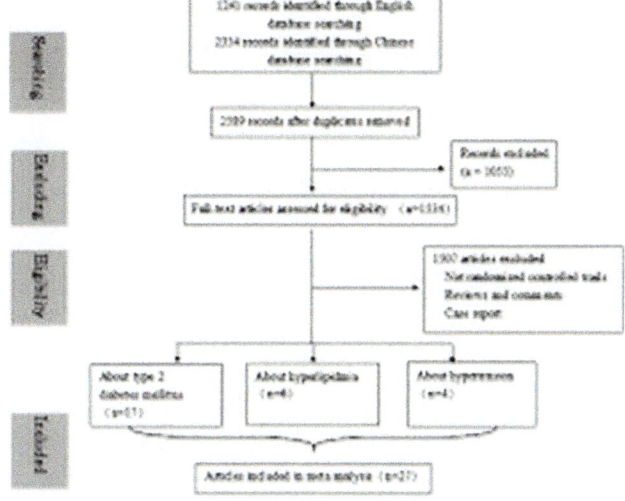

FIGURE 6.8   Treating type 2 diabetes. (From Lan et al. (2015).)

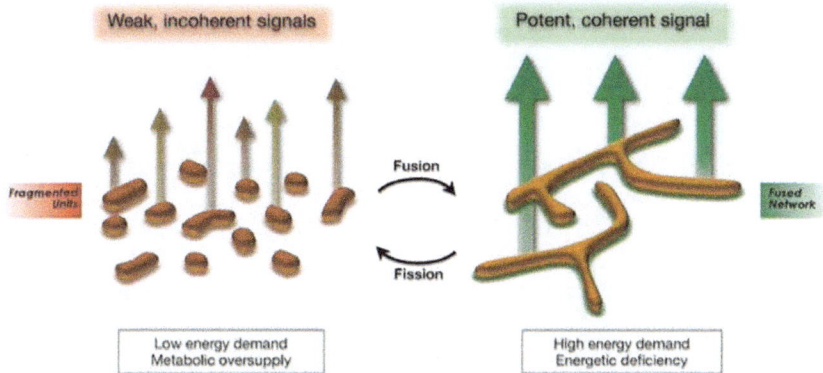

FIGURE 6.9   Mitochondria: starving to reach quorum? (From Picard and Burelle (2012).)

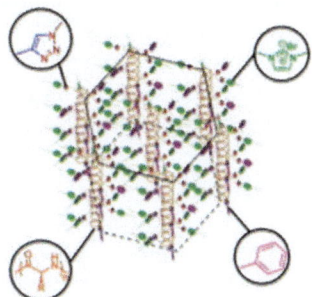

FIGURE 6.10   Thermotropic liquid crystalline polypeptide bearing imidazolium and *p*-tolyl groups. (From Hu et al. (2015).)

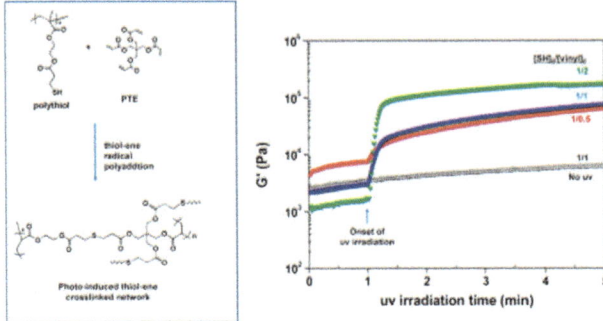

FIGURE 6.11 New photo-induced thiol-ene crosslinked films based on linear methacrylate copolymer polythiols. (From Zhang et al. (2013b).)

composed of linear methacrylate copolymer polythiols are reported, exhibiting rapid cure, a uniform network, and enhanced mechanical properties for surface coatings applications." (Zhang et al., 2013a). What is the quality of the image and the highlight?

4. *Highlights*:

(a) Which of the following highlights adds value versus what is already in the title "Synthesis and solid-state properties of thermotropic liquid crystalline polypeptide bearing imidazolium and *p*-tolyl groups" (Hu et al., 2015):

- We prepared thermotropic liquid crystalline polypeptides bearing imidazolium and *p*-tolyl groups.
- We examined the solid-state structures and thermal properties of the resulting polypeptides.
- Increasing *p*-tolyl contents give rise to more stable liquid crystalline phase and higher glass transition temperatures.

(b) Graphical abstracts in the *Journal of Ecology* often have photos and highlight text to the left of the photo. The image gets bigger when you click on it so the clarity is good. The highlight for Figure 6.12 reads "Our results demonstrate that the strength of density-dependent seedling survival can vary between seasons and among species in tropical forests. Future research is needed to assess the underlying mechanisms of this temporal and interspecific variation in NDD and its consequences for species coexistence and community composition." What is the quality of the image and the highlight?

5. *Introduction*:

(a) Identify deficiencies in the following sentence: "In the effort to reduce greenhouse gas emissions and to limit the gradient of carbon concentration increase in the atmosphere, CO_2 capture and storage (CCS) is

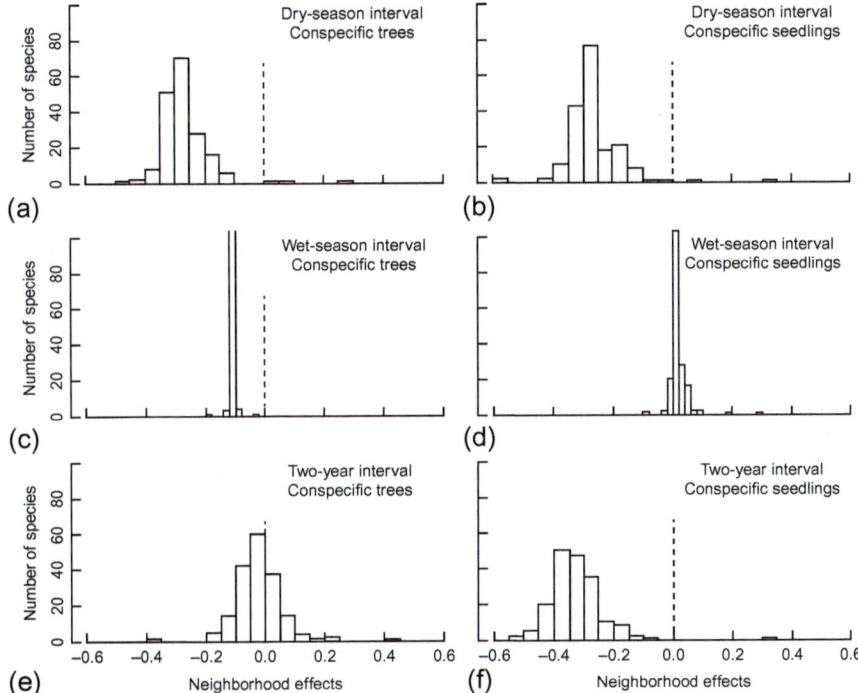

FIGURE 6.12 Seasonal differentiation in density-dependent seedling survival in a tropical rain forest. (From Lin et al. (2012).)

receiving increasing attention in many countries worldwide" (Pehnt and Henkel, 2009).

(b) Identify deficiencies in the following sentence: "Carbon dioxide reforming of methane to produce synthesis gas, i.e., a mixture of carbon monoxide and hydrogen ($CO_2 + CH_4 \rightarrow 2CO + 2H_2$; $\Delta H^o_{298K} = 247.0\,\text{kJ}\,\text{mol}^{-1}$) has attracted substantial interest (1-4)" (Bitter et al., 1997).

(c) Identify deficiencies in the following sentence: "Diabetes, hyperlipidemia and hypertension are the clinical syndromes caused by the compounding genetic and environmental factors. They are common diseases, frequently occurring on a global scale" (Lan et al., 2015).

(d) This book has 372 references. The age of the references (the difference between 2014 and the year the citation was published) follows a Weibull distribution with a modulus of 0.8, and 10% are older than 25 years.

   i. Derive the value of the scaling parameter, θ.

   ii. What is t_{50} of the age of the references?

   iii. How many references were written in 2012, 2013, and 2014?

iv. What is the difference in the age distribution if the modulus is greater than 1? If the modulus is less than 1?

v. Is this age distribution adequate for a journal article?

(e) Calculate the modulus and scaling parameter for the age distribution of the references in Zhou et al. (2011). How would you characterize the age distribution of the references?

TABLE 5e Most of the papers of Zhou et al.'s (2011) cite were less than 3 years old

2003	2003	2010	2009	2010	2010
2005	2010	2008	2010	2010	2009
2005	2000	1993	2010	2010	
2007	2010	1994	2010	2010	
2004	2009	2003	2010	2009	
2008	2008	2010	2010	2009	
2005	1994	2010	2010	2010	

(f) Calculate the modulus and scaling parameter for the references in Leucht et al. (2009). How would you characterize the age distribution of these references?

TABLE 5f Most of the papers of Leucht et al.'s (2009) cites were more than 3 years old

2005	2006	2007	1999	1999	2003
2000	2007	2007	2002	1969	
2003	2003	2005	1991	1987	
2006	1986	2005	1990	1993	
2005	2004	2003	1990	2007	
1997	2003	1988	2008	2005	
2005	2002	2008	2006	2003	
2006	2006	2005	2006	2003	
2007	1997	2005	2007	2006	
1997	1989	1997	2007	2005	

6. *Methods*: Correct the following sentences:

(a) The urea homogeneous[2,7,18–21] method was used to prepare spherical particles of Y_2O_3:Eu phosphor. (Martinez-Rubio et al., 2000)

(b) The CL measurements were carried out using a high-vacuum (5×10^{-7} Torr) chamber and a Kimbell Physics Inc. (Walton) model EGPS-7 electron gun. (Martinez-Rubio et al., 2000)

(c) Enhanced surveillance was implemented in the United States for human infection with influenza A viruses that could not be subtyped. (Dawood et al., 2009)

(d) Two assay systems were used to measure the antiviral activities of the PQ. (Hudson et al., 1997)

(e) X-ray diffraction (XRD) study of all the samples was carried out at room temperature on a D/Max 2500V/PC (Rigaku Corporation, Tokyo, Japan). The crystallite size of pure LLDPE and DA-G/LLDPE nanocomposites was carried out by using the Scherrer formula (Kuila et al., 2012)

$$D_p = \frac{0.89\lambda}{\beta \cos\theta}.$$

7. *Conclusions*: Identify how you could improve these conclusions:

(a) Because of its geological and climatic characteristics, Calabria is very prone to WSE. This study, by comparing different scenarios, highlights the effectiveness of some antierosive practices and the likely increase in intensity of erosive processes as a consequence of wildfires in woodlands. (Terranova et al., 2009)

(b) This review paper focused on aspects of precipitation research relating to measurements and applications. Research on precipitation can definitively advance our ability to predict the water cycle mechanisms contributing to extreme events (floods, droughts). Arguably, improving the use of observing systems at global and local scale, the understanding of the involved physical processes and the integration of measurements in atmospheric and hydrologic models will greatly advance our capability to make better quantitative precipitation forecasts with consequential improvements in hydrologic forecasting and water resources management. It is pertinent, at this point, to note the role... (Michaelides et al., 2009)

(c) This paper describes the Watch Forcing Data (WFD) created at half-degree resolution for the purpose of driving LSMs and GHMs through the twentieth century. For the period 1958-2001 the WFD can be considered to provide a good representation of real meteorological events, synoptic activity, seasonal cycles, and climate trends. (Weedon et al., 2011)

(d) The products of banana production, which are consumed globally, make this crop an object of common interest. Characterisation must be undertaken of its content and concentration of useful biomolecules. The variety of harvesting methods, post-harvesting treatments and consumption methods used throughout the world make bananas and plantain a useful model for investigating quality and its development in fruits and vegetables, through the processes from field to table. Consumption forms (fresh or processed) should also be taken into account, in terms of quality of products consumed, and the dynamics of antioxidant and other bioactive compounds. (Aurore et al., 2009)

(e) This study shows that LANDSAT satellite images can be used for improving the mapping of soil salinity. The best band combinations for estimating soil salinity with different crops are as follows: alfalfa (red, near infrared, and NDVI); cantaloupe (blue and green); corn (near, thermal, and NDVI); and wheat (blue and thermal). The OK model performed the best, the RK model performed second best, and the CK model had the worst performance. The better performance of OK over RK may be attributed to the fact that autocorrelation among soil salinity data are higher than cross correlation between soil salinity and the LANDSAT bands.… (Elfeiry and Garcia, 2010)

(f) A simple MAE procedure, followed by a SPE purification step prior to HPLC analysis and spectrofluorometric detection, was optimised for PAH determination in smoked meat samples. When a Carboflon bar is used to heat the apolar solvent, hexane can be successfully used as extraction solvent for MAE. The optimised conditions for PAH extraction from lyophilised smoked meat samples were the following: 2 g of sample extracted with 20 mL of hexane at 115 °C for 15 min. With the exception of Pitina samples, other commercial products presented a very low PAH load. (Purcaro et al., 2009)

(g) The results show that all models simulate much of the long-term statistical characteristics and the oxygen dependent dynamics of the biogeochemical cycles of nitrogen and phosphorus. There are differences between the model results and possible underlying reasons have been discussed. One important finding from the present analysis is that none of the models is significantly better or worse than the others and that the models support each other to give an added value for the ensemble mean assessment that is shown to be as good as or even better than the results from any of the individual models. Because the models show different performances and statistics of different variables the good ensemble mean assessment performed at representative stations for the Baltic Sea supports the use of ensembles for climate change studies.… (Eilola et al., 2011)

8. *Acknowledgments*:

(a) If you conceive an experiment and collect data to support it, does that qualify you as a coauthor?

(b) If you have applied for a patent and then publish a paper based on the results, what must you disclose?

(c) If you are a director of a company and a professor, do you need to declare the dual role? What if you only consult for the company?

(d) The paper entitled "The effect of intensive treatment of diabetes on the development and progression of long-term complications in insulin-dependent diabetes mellitus" (Shamoon et al., 1993) has 681 coauthors. How many coauthors have the same last name and initial(s)?

Chapter 7

Presentations They Will Remember

A great presentation, like a great manuscript, is **precise, concise, vigorous,** and **straightforward**. It isn't a mystery novel or stand-up comedy. You will lose the audience if you save the punch line until the end. Begin by summarizing the important contributions of your research (even with the first slide). Metadiscourse is appropriate for presentations; you want to repeat your message several times during the talk. At the end, repeat what you said at the beginning: An outstanding presentation maintains the audience's attention from the title slide to the last question. You fail when people start checking their phones, yawning, talking, or working on their computers (not to mention those who leave midway through the talk).

What you say, how you say it, and what you project on the screen is as important as the content. When you attend a presentation, do you focus on what presenters say or on the slides that they project on the screen? What is the relative importance of slides versus speaking? Politicians transmit their ideas with words; researchers communicate their discoveries with words, images, graphs, tables, animations and videos. The content of your presentation is the most important element but how you express yourself and what you project will determine if people will understand your research.

Communicate Science Papers, Presentations, and Posters Effectively. http://dx.doi.org/10.1016/B978-0-12-801500-1.00007-3

Slides
- are synoptic,
- help pace the talk,
- substantiate what you say, and
- stimulate interest.

You have to strike a balance between speaking and showing. Slides complement what you say so keep the text succinct. People can't follow what you say and simultaneously digest pages of dense text or multiple rows and columns of data. We are capable of doing two things at the same time—walking and chewing gum, for example. However, your brain is incapable of devoting equal time to multiple stimuli during lectures, conversations, and presentations (Medina, 2008). In TED talks, slides play a secondary role to reinforce the speakers message. Too often in scientific presentations, speakers repeat everything on the slide. Slides are props, the dessert, eye candy, they are not the main course. That is not to say that slides are unimportant. Au contraire, people remember pictures six times longer than either text or sound (Medina, 2008). But, images support what you say, they don't duplicate it. People's ability to recall increases by adding multi-media content to hearing and seeing—animation, video, and touch. Bill Gates released mosquitoes in the auditorium during his talk on malaria (www.ted.com/talks/bill_gates_unplugged), which was more powerful than videos or animations. (John Medina, Brain Rules, Pear Presse, Seattle, 2008—cited by Gallo et al., 2014).

7.1   CONTENT

Poor slides equal poor presentation. The best presentations come from people with the best slides. Logically, the two are related: people who spend time preparing slides spend time organizing their talk, reviewing it in their mind, and thereby rehearsing it. The first principle of speaking is to be organized. The second is to speak well, which comes with practice and repetition.

Presentations should have no more slides than the number of minutes for the talk, and no less than half. For a 15 min presentation, target 9 to 13 slides. For a 30 min presentation, 20 to 25 slides are enough. However, don't overcharge them. Gallo (2014) stated that PowerPoint slides average 40 words per slide. However, he suggested that you should target a total of 40 words in the first 10 slides. Moreover, he recommends that one theme per slide is appropriate.

Title Slide

Constrain the title to no more than two lines, preferably one line. Consider shortening the title you wrote for the proceedings. Remember, slides complement what you say. If right from the beginning you lose the people in the audience with a title that is too long you might not get them back.

FIGURE 7.1   This title slide carries all pertinent information and as well as four major topics of the talk.

Other pertinent information on the title slide includes authors, affiliations, date, conference venue, and an image representing the talk—a photo of the product, experiment, or process. For multi-institutional papers with more than five authors, put most of the coauthors in the acknowledgments and cite a few key authors on the first page. The audience is unable to register so many people, and it crowds the title slide with irrelevant information.

The title slide can also serve as the outline of the talk (Figure 7.1). The topics are not IMRAD headings, but rather keywords that represent the work.

Summary Slide—Second Slide

If you don't summarize your work on the title page, the second slide is the most important slide, where you

- describe the context,
- state the results, and
- explain why they are significant.

This slide is equivalent to a graphical abstract and to highlights—(not objectives, hypotheses, or purposes). Don't surprise everyone at the end of your talk with your brilliant contribution. State it clearly at the beginning.

Besides images and text, integrate a demonstrative table of contents: avoid boilerplate words such as IMRAD (Figure 7.2). Name the pertinent sections of your talk (Table 7.1).

Include a tagline on every slide. When you transition from one to another, add a square box to highlight exactly where you are (Figure 7.3). This presentation describes a novel impact resistant polymer for cell phones. The tagline helps the audience follow the talk, and separates the title from the content. You can

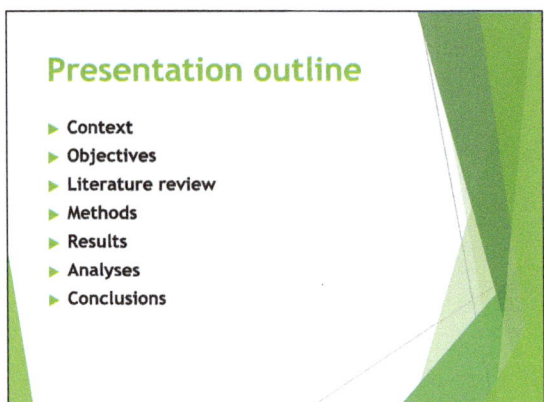

FIGURE 7.2   Replace outlines containing standard section titles with text that represents the work. A dedicated slide for an outline is unnecessary.

TABLE 7.1 Replace Outlines Containing Standard Section Titles With Text That Represents the Work

Title	Improving Smartphone Impact Resistance	Alternative Acrylic Acid Technology
Introduction	Product targets	Current technology
Methods	Market opportunity	Catalysts: VPO versus MoVSb
Results	Technology platform	Reactor technologies
Discussion	Innovation	Economics
Conclusions	Implementation strategies	

3/14	Targets	Market	Technology	Innovation	Economics

5/14	Targets	Market	Technology	Innovation	Economics

7/14	Targets	Market	Technology	Innovation	Economics

FIGURE 7.3   Rectangles around text in the tagline indicate where you are in the presentation. This helps people who come late to the talk or those who become distracted.

place the tagline at the bottom, on the side, or in a corner. A font size of 18 pt is appropriate. Images are more powerful than text (Figure 7.4): we replaced *Targets* with a dart board, *Market* with an image of the world, *Technology* with a molecule, *Innovation* with a light bulb, and *Economics* with a dollar sign. Rather than framing the section with a rectangle to indicate where you are, brighten the

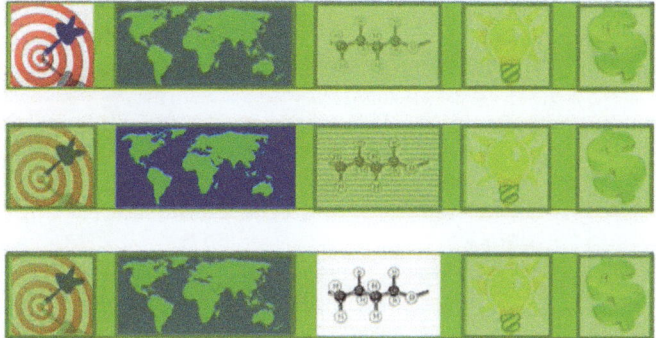

FIGURE 7.4   Rather than text and rectangles, images can represent the themes of the presentation. Dim all tagline images except for the section you are addressing.

FIGURE 7.5   Replace outlines containing standard section titles with text that represents the work. Bullets are better than sentences, but hedging introduces irrelevant text.

image that corresponds to where you are, and dim the rest. Additional slides demarcating the transition to a new section are unnecessary.

Figure 7.5 is better than Figure 7.2 because it summarizes the salient features of the talk, but Figure 7.6 is even better—hedging is gone, so there are fewer words and more white space. The hedges are *market projection*, *might be*, *performance improvement*, *could be*, and *range from*. The project costs $6 million and improves performance by 25%. It doesn't matter if the target was 50%. The information in this slide is pertinent to the facts. The audience understands the context of the talk: it relates to expected performance and cost. Timid researchers hedge often, but that doesn't make it better, it just adds meaningless words.

Figure 7.5 has twice as many words as Figure 7.6. The latter's title is pertinent and occupies less space. The page number and the total number of pages are at the bottom right corner. We replaced *performance improvement*, which is vague,

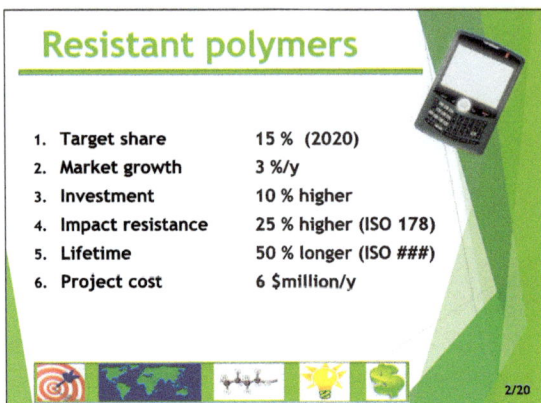

FIGURE 7.6   Place taglines on the sides, along the bottom, in the corners, or below the title to separate it from the content.

with *impact resistance* together with a reference to the ISO standard. The image of a smartphone emphasizes that the application is for impact resistant polymers, which you would mention. You could replace the numbers and words in the left column with images.

Background

After the summary slide, take some time (let's say 3 min to 4 min of a 15 min talk) to describe the context of the problem and your research—three to four slides. While we called this section *background*, choose a word that is related to the subject of the talk; for example,

- current (alternative) technologies,
- challenges,
- experimental (identify specialized equipments or techniques),
- performance criteria (e.g., ISO standards),
- incentives.

Highlight the originality of your research, and make sure you mention previous work. Delegates have been known to shame speakers by asking them publicly if they thought that they were the only ones working in the field. Remember to favor images over text—combine images with text for an experimental apparatus. Don't fear that they won't believe you because you haven't reported all of the details. Doctoral theses require proof (and Ph.D. students have more time during their defense). Audiences assume that you have followed adequate protocols to derive your results. For controversial subjects, some people will be interested in the minutiae, but that doesn't mean everyone in the room must suffer the details. Offer to talk about them after your presentation.

Results

For a 15 min presentation, spend 4 min on the introduction and methods and a couple of minutes on the conclusions; dedicate the rest of the time to the results. If the audience is familiar with the topic (specialized conferences), spend less time introducing it and more time on the details, but avoid professional narcissism: Assume the audience is competent but has never worked in your field. Minimize jargon and define terms that are key to understanding what you are saying. Sometimes results consist of a single data point. The novelty relates to how you got it or how you extrapolate it to various situations or systems. In this case, concentrate more on the methods and discussion and less on the result.

Visual aids—graphs, tables, videos, and animations—support what you say. People listen to what you say but switch their focus between you and the slides. Design your slides with plenty of white space (or a monochrome background color), otherwise they will find it harder to listen to you and assimilate the images.

Discussion/Conclusions

The conclusions should not be a shocking revelation to the audience. It repeats the message you set out to give from the beginning. Highlight unfinished work, and recommend possible avenues of research. Alternatively, use the summary slide that you presented at the beginning of the talk as the conclusions slide. Remind the attendees of the importance of the work—economic incentive, scientific contribution. Resist introducing new data or alternative interpretations of what you presented, but consider identifying the implications with respect to other technical fields and the opportunities. Charts and images are acceptable.

Acknowledgments

Coauthors' names are on the title slide as well as in the conference proceedings. You may itemize individual contributions at this point: Prof. X conducted the XPS analysis, postdoctoral fellow Y compiled the database, Ph.D. student Z did the experiments, etc. However, this information is irrelevant to the subject; the team did the work. Show a picture of the research team, including individuals who helped but that were not directly responsible for any part of the presentation. Acknowledge organizations that funded the work as well as all institutions involved. Displaying the logos is sufficient.

7.2   STYLE—FORMAT

Remember that because you can read it on your screen doesn't mean people at the back of the conference room can read it. Large font, big images, contrast, and colors make the difference. Keep slides simple. Animating text and images is sometimes desirable to emphasize important concepts. Animating too many slides (more than one in three) becomes annoying. Don't exaggerate the special effects (Tavares and Virgilio, 2014), unless you work for Lucasfilm.

Text

Unless you quote someone, write sentence fragments of six words or less. More than that and you will lose some of the audience members as they read what you wrote instead of listening to what you say. They read faster than you talk, but they can't do both and check their smartphones.

The minimum text font size is 20 pt (18 pt for some fonts). Anything smaller than that clutters the page and is hard to read. Avoid walls of text and clichés such as *This is an eye test.* Make it clear for the people at the back of the room. If you present well, even these people might stay instead of escaping out the back of the room halfway through your talk.

Too little text is rarely annoying, but what can look juvenile is a title and a single sentence in the body (Figure 7.7). A font size of 54 pt and 44 pt for the title and the body, respectively, is too big. The authors could have combined the text into a single slide.

Use colors and bold typeface to highlight important points or words. (Minimize italic typeface.) Sans serif fonts such as Beamer, Tahoma, Arial, Helvetica, Calibri, and Lucida Sans are better than serif fonts such as Times New Roman, Book Antiqua, Century Schoolbook, and Georgia. Restrict text in the slides to no more than three font sizes.

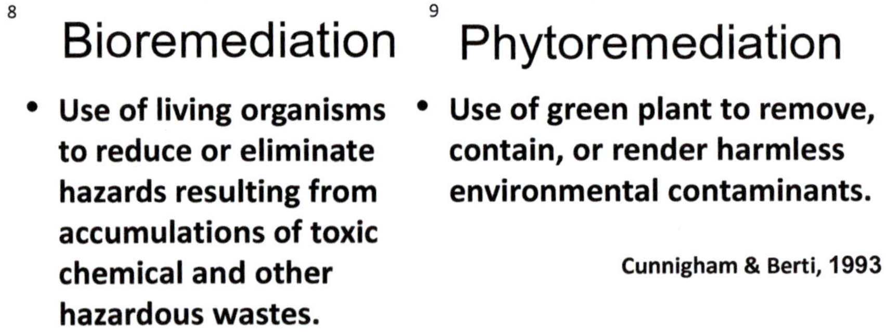

[8]

Bioremediation

[9]

Phytoremediation

- **Use of living organisms to reduce or eliminate hazards resulting from accumulations of toxic chemical and other hazardous wastes.**

- **Use of green plant to remove, contain, or render harmless environmental contaminants.**

Cunnigham & Berti, 1993

FIGURE 7.7   Adjacent slides with a 54 pt title and 44 pt text (quoted by Davis, 1997).

Title Slide

Choose a font size for the title from 28 pt to 44 pt. The font size of the list of authors should be smaller—say, 24 pt—and all other information should be smaller still, maybe 20 pt but no less than 18 pt. Colors and images are appropriate. For short presentations, adapt the title slide to present the context of your talk.

Slide Titles

As with the title slide, short titles are better than long titles. If you choose to have a title, left-adjust it and maintain the same position, font type, and size for all slides with titles. Don't reduce the font size to fit the title in one line. Rather, reduce the number of words.

Lists

Bullet lists become boring; break the monotony by adding images. In fact, images by themselves are best. Figure 7.8 shows a leaf and the opportunities and constraints of developing agricultural programs in Central Africa. Initially, the brown twig has no leaves, but with each successive mouse click a leaf appears with text—three words or fewer. Keep the maximum number of items in a list to about six. Concentrate on the important points.

In the project schedule example (Figure 7.9), the background slide has just one large sphere at the tip of the arrow. With each click, the text in the large sphere changes, and a smaller sphere appears along the arrow with the text from the previous sphere. The text gets smaller the further the spheres are from the tip of the arrow. Like the leaf example, these graphical elements keep the audience's attention from wandering. You focus their attention on what you're saying, and the text reinforces the message. Some people will stray from one sphere to another if you present all of them at the same time.

Animations such as these restrict people from meandering from one bullet point or image to another. This technique is effective when the slide has more than three bullet points. Figure 7.10 combines animation, text, and images.

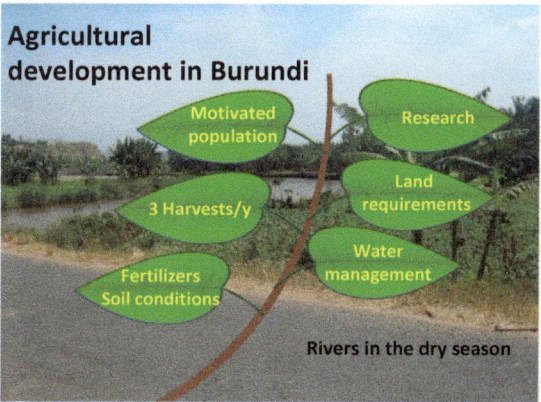

FIGURE 7.8 Agricultural qualities are displayed on the leaves. With each mouse click a leaf appears. (From Karirekinyana and Patience (2010).)

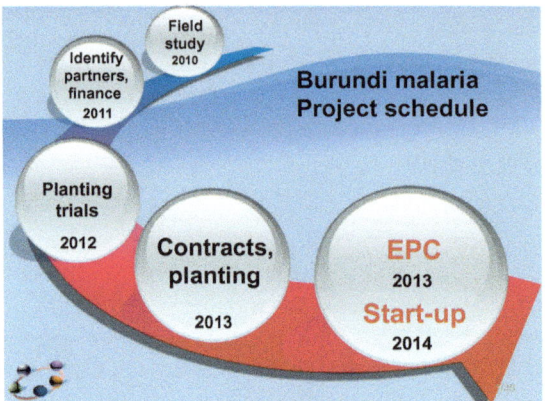

FIGURE 7.9   An animated slide to follow the steps of a development project: with each mouse click, text in the large circle changes and a new smaller circle appears with the text from the previous circle. The colors reinforce the arrow's direction. (From Karirekinyana and Patience (2010).)

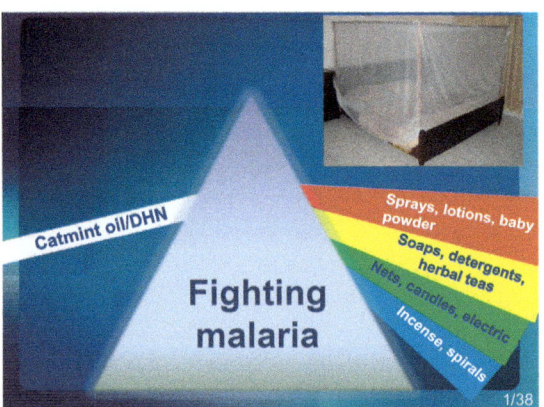

FIGURE 7.10   Fight malaria with catnip (an essential oil) integrated in lotions, sprays, candles, and bed nets. This slide combines a list, images, and animation—on every mouse click the image in the top right corner changes, and a new light beam appears that corresponds to the image. (From Karirekinyana and Patience (2010).)

The initial slide is a prism entitled "Fighting Malaria" with a monochromatic light beam that represents the solution—catmint oil as a substitute for standard mosquito repellents. With the next mouse click, an image appears in the top right corner together with text in a red light beam. With subsequent mouse clicks, more light beams appear with the corresponding images in the corner until the very last light beam with a photo of a bed net.

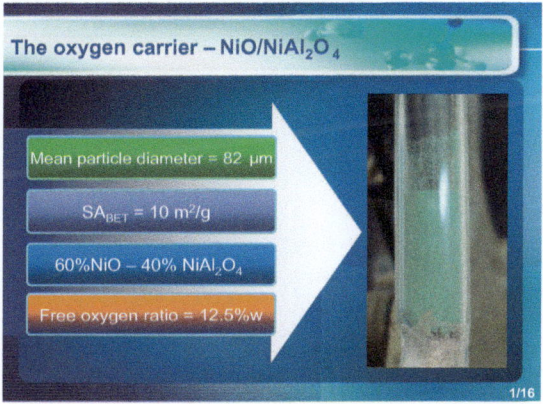

FIGURE 7.11   Property data table: it combines numerical data and a video to the right of NiO powder bubbling in a microreactor.

Tables

Science articles have few tables. You should have even fewer tables in your presentation. Integrate data with images, or report them as graphs. If you lack time to do this, then follow the guidelines in Chapter 5 to make tables correctly. White space is good, so don't worry if there is not much data and there is a lot of space. Figure 7.11 combines an image of a microreactor and a property table with four data points. Animating the text is unnecessary.

Keep the font size in tables the same as for the rest of the text. If you can't fit all of the data into a table, then simplify the table. Too much data puts people to sleep. We said that six bullet points with six words per bullet point was the maximum to expect people to assimilate in a slide of text. That is a good rule of thumb for tables—including the column and row headers, 6 rows × 6 columns of data. It could also be too much. You will not address every data point individually, so highlight the important points in bold or color, or increase the font size. Animate the table and change the font while you describe the individual points. You may also frame an individual column or row for emphasis. Intersecting frames or shades help focus the audience's attention on significant data. Another technique is to dim or gray out unimportant data points (rows and columns) (Table 7.2).

Table columns and row headings should have units.

Graphs and Images

Day and Gastel (2011) contend that if it takes more than 4 s to understand a slide, it has too much information. According to this criterion, slides should have fewer than 16 words: people read at 240 words per minute, and so it takes 4 s to read

TABLE 7.2 Highlighted words are ionized verbs

Word	Web of Science $N/10^3$	%	Nature, Science N	%
Studies	1853	4.8	3762	2.6
Use	1834	4.7	283	0.2
Analysis	1288	3.3	173	0.1
New	993	2.6	1012	1.7
Application	595	1.5	105	0.1
Development	592	1.5	561	0.4
Characterization	505	1.3	1102	0.8
Novel	448	1.2	151	0.1
Performance	428	1.1	161	0.1
Assessment	301	0.8	197	0.1
Investigation	246	0.6	109	0.1
Enhanced	231	0.6	333	0.2
Advances	170	0.4	1166	0.8
Observation	124	0.3	154	0.1
Modification	78	0.2	642	0.4
Total	9686	24.9	9911	6.9

16 words. Slides should be intuitive, with more graphs and images than text, but even cartoons take more than 4 s to understand.

Help the audience grasp the essence of the slides by dividing them into regions. Slides are 250 mm × 190 mm, so break them up into 3 × 2   80 mm squares. This coincides with the maximum of six bullet points that we recommend for a slide. Coincidentally, 80 mm is close to the standard journal column width. Having six regions doesn't mean you have to fill them all. Remember that empty space is a powerful way to emphasize important concepts. Combine regions vertically or horizontally to accommodate larger images in these directions (Figure 7.12). For very important images, combine four adjacent regions to make a square. We have delimited the region with bubbles and placed text inside and outside the bubble (Figure 7.12). Optionally, add contour lines around the bubbles and text.

Ya-Huei (Cathy) Chin masters this format (Tu and Chin, 2014). Professor Chin changes some images inside the bubbles while advancing from one slide to the next. This strategy gives continuity to the talk. It directs the audience's attention to new images incrementally, and allows more time to assimilate the context of the rest of the slide.

We reproduce this effect in Figures 7.13 and 7.14: A picture of an attrition mill occupies two regions to the far right. In the bottom left, a block of text lists the operating conditions. A scanning electron microscope image of a mixture of salt and the catalyst vanadyl pyrophosphate (VPP) straddles the middle of the page. Images and text occupy four of the six regions. On the first mouse click, a photo of a furnace appears in the bottom middle region. Simultaneously, red text

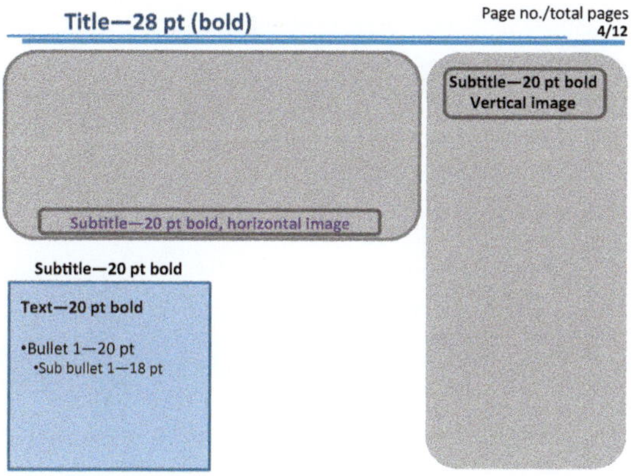

FIGURE 7.12 Master layout: Divide the space into six regions. Here, two regions are joined vertically to the right and two are joined horizontally below the title. An empty space defines the bottom middle region.

appears in the block of text describing the upper operating temperature of the furnace. The page numbers increase by one (from 5 to 6) with the next mouse click, and images of attrited catalyst and salt replace the block of text and the photo of the furnace.

With the following mouse click, the page number advances from 6 to 7, and two images related to CaO-CuO replace the three images of VPP and salt. The theme is the same as for the previous slide—attrition—but it is for a different system. The next mouse click increases the slide number by one, and a graph of the attrition resistance of three systems versus gas velocity fills the bottom middle region. A video of the attrition mill replaces its photo. Finally, on the next mouse click, a regression line appears on the graph that characterizes the relationship between attrition and gas velocity (we don't increase the page number by one).

Sequencing images such as this helps the audience members follow the presentation because it doesn't overwhelm them with new information with each slide.

Graphs (*x-y* plots, bar charts, pie charts, etc.), photos, and schematic diagrams are the commonest images. Chapter 4 outlines best practices for graphs in papers and presentations. Plot dimensions for papers are too faint for presentations (Figure 7.15). For the same plot dimensions (*x-y* length in millimeters), lines must be double the weight for a presentation or poster compared with a paper. Text and symbols should also be at least double the size of that of papers. Bold text is also attractive, and a smaller font is acceptable (Figure 7.15).

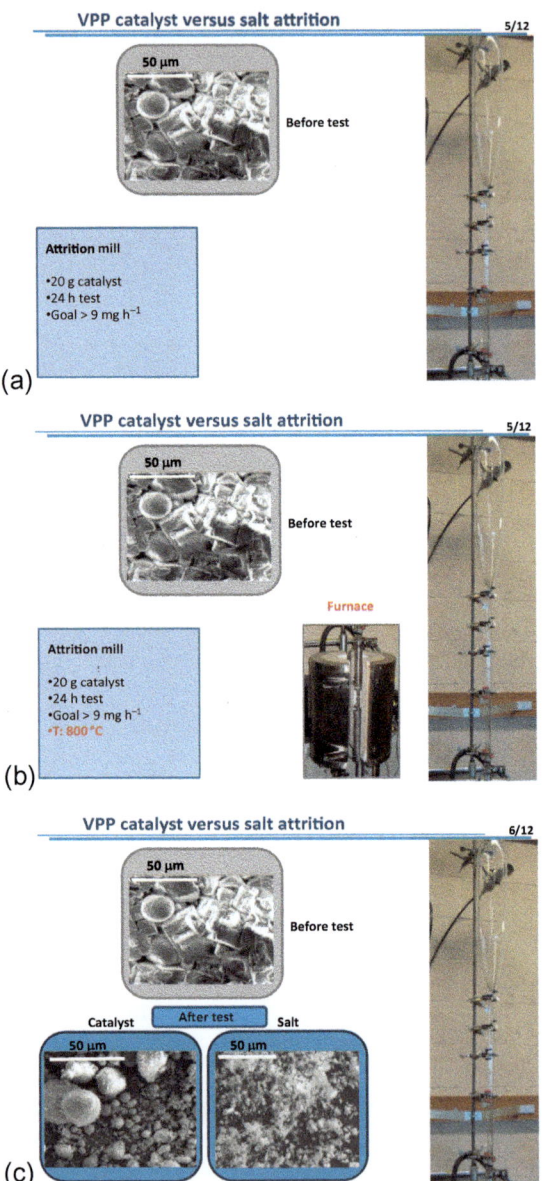

FIGURE 7.13 (a) A photo of a catalyst and salt replaces the horizontal bubble, and a photo of the attrition mill replaces the vertical bubble. A text box summarizes the experimental conditions. (b) A photo of the furnace fills the white space next to the attrition mill. The text box shows the maximum temperature. (c) Images of the attrited catalyst and salt replace the furnace and text. (a) 1st slide; (b) 1st mouse click; (c) 2nd mouse click.

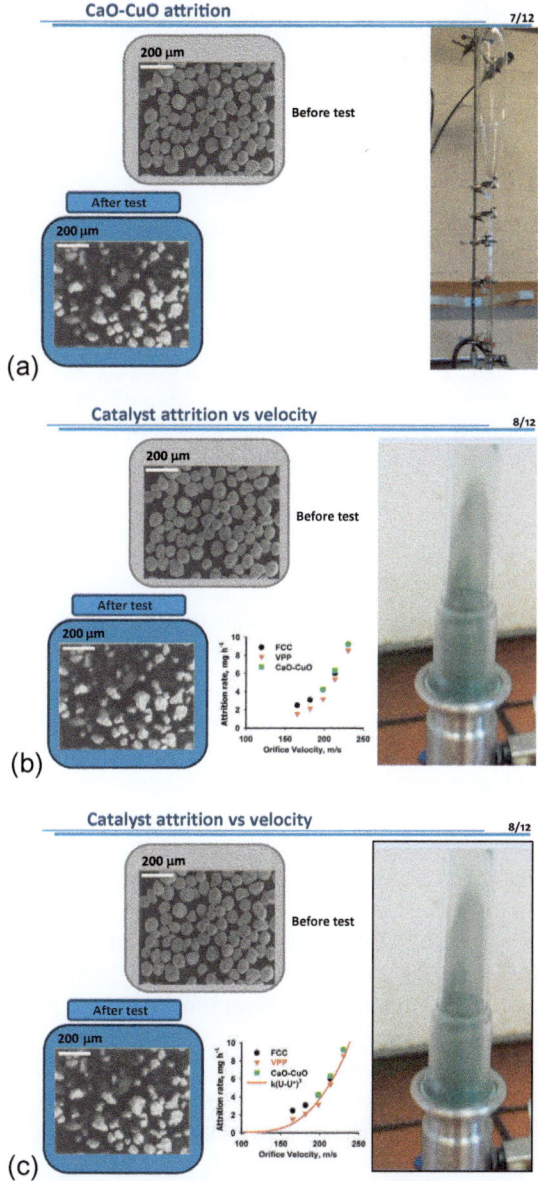

FIGURE 7.14 (a) CaO-CuO images replace the salt and VPP photos. (b) A graph of the attrition rate of three catalysts versus velocity fills the bottom middle region. A video of the powder fountain that forms at the bottom replaces the photo of the attrition mill. (c) A regression line with the model appears on the graph. (a) 3rd mouse click; (b) 4th mouse click; (c) 5th mouse click.

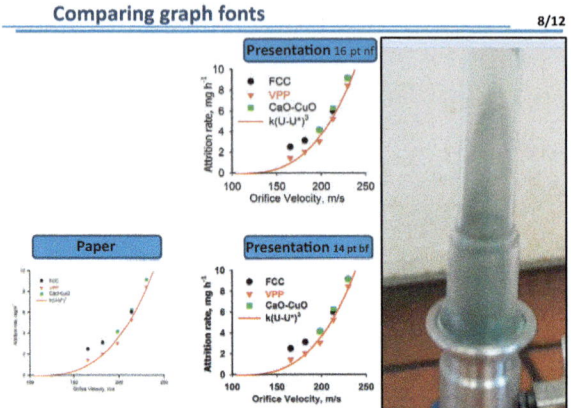

FIGURE 7.15 Standard text for a 55 mm high plot for a paper is too small for a presentation. Fourteen point bold font (bf) is sufficiently large and is in harmony with the bold font of the titles. The 16 pt normal font (nf) text is equally legible.

Neither tables nor figures carry labels such as Figure # or Table #. Minimize text. Unlike in papers, use symbols for the variables (T, °C instead of temperature; t, s instead of time) to make the axes' labels easier to read. Rather than explicative captions, write short titles at the top of figures: "Temperature versus time," for example. This way you won't confuse it with the x-axis title. Let the audience hear you describe the figure with words that would have gone in a caption. Seldom should you read text on the graphs or images.

Figure 7.16 combines graphs and experimental operations. In Figure 7.16(a), methane, represented by a green line, goes to the reactor. The graph to the right shows that the methane concentration approaches an asymptote at 40 s. In the next slide (Figure 7.16(b)), oxygen goes to the reactor—the red lines in the figure to the left. The graph to the right represents its concentration with time. Finally, in the third slide (Figure 7.16(c)), argon (magenta) purges oxygen from the reactor.

Gradient fill with multiple colors from top to bottom or from side to side is artistic. The best gradient fill is to start dark at the top and progress to white or a very pale color at the bottom (Figure 7.16). Dark backgrounds can be attractive, but most graphs and other documents have white backgrounds. Adapting figures to a dark background adds work. An occasional bright red or black background in an otherwise white-background presentation can keep your audience awake. (So can distributing coffee before your talk.)

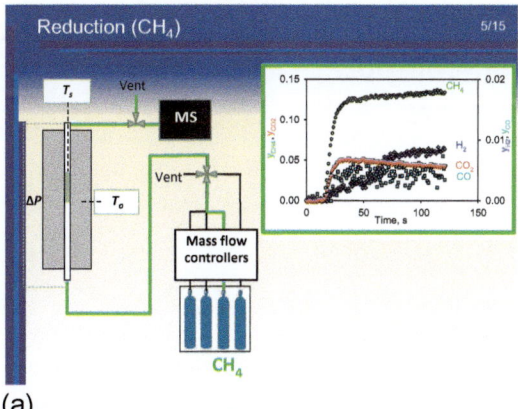

(a)

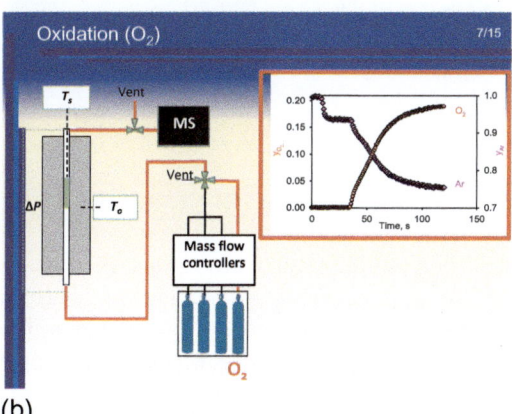

(b)

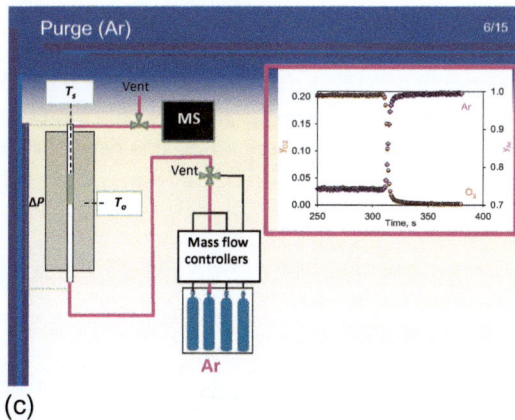

(c)

FIGURE 7.16 These slides combine graphs with an experimental sequence. The line colors represent the gas compounds, and we reproduce these colors in the *x-y* plots. (a) CH_4, in green, passes from the gas cylinder to the reactor and then to the exhaust and the mass spectrometer (MS). (b) O_2, in red, passes through the reactor. (c) Ar, in magenta, purges the reactor. (a) Reduction; (b) Oxidation; (c) Purge. (From Perreault et al. (2012).)

Wolpaw, Birbaumer, McFarland, et al. Clinical Nerophysiology, 2002 113(6) 767.

FIGURE 7.17   Abbreviate references in a footnote at the bottom of the page or include the header of the article. (From Wolpaw et al. (2002).)

Footnotes

It's mandatory to put the slide number and the total number of slides in a corner (Figure 7.6) or at the beginning or end of the tagline (Figures 7.3 and 7.5). People want to know how many slides you have, and where you are in the presentation.

Whereas we insist that the font size be 20 pt or greater, footnotes should be smaller: 14 pt to 16 pt—the information is repeated and is less important than the rest of the slide. Partition the footnote with a horizontal line when you want to add references. Abbreviate it with the surnames of the first three authors. The font size depends on the face type: 12 pt to 14 pt is good for Arial, but 14 pt to 16 pt is better for Calibri. Alternatively, copy the journal header and include it on the slide (Figure 7.17). You can add the date to the footnote as well as the conference name and your name (so people don't forget it). Some have the institutions' logos at the bottom of every slide. It is better to maximize the slide space for your data, so if you choose to add a footer, it shouldn't occupy more than 5% of the slide.

7.3   COLORS

To save time when you start a new presentation

- get your institution's presentation template (Tavares and Virgilio, 2014),
- adapt predefined templates from graphics packages online,
- keep slides to four colors or fewer (besides black and white),

- create a color palette with online programs (e.g., paletton.com),
- copy one you have already made or borrow someone else's, and
- choose white or a very pale color for the background.

Primary colors for pigments (paint) are cyan, magenta, yellow, and black (CMYK), whereas for light (computers, television), the monochromatic primary colors are red (700 nm), green (546 nm), and blue (436 nm)—RGB. On the RGB color wheel, the secondary colors are yellow, cyan, and magenta, which are equal mixtures of red and green, green and blue, and blue and red (Figure 7.18). Tertiary colors are mixtures of adjacent primary and secondary colors, and together with primary and secondary colors comprise the 12 hues. Amaranth is a color with a red hue, and rust has an orange hue. Red, orange, and yellow are considered warm colors, while green, blue, purple, and violet are considered cold colors. *Tint* refers to colors becoming paler (adding white pigment for paint), and *shades* refer to colors becoming darker (adding black pigment). Red, orange, blue, and violet are good colors for titles on a white background. Greens, yellow, and pink need shading for text and titles. For bullet points, black on white is good, with one or two words or phrases in color.

Rather than a color wheel, you can create a color palette from an RGB contour plot (e.g., PowerPoint—Figure 7.19). The total number of color combinations $\{x, y, z\}$ in a 24-bit pixel is $2^{24} = 256 \times 256 \times 256 = 16\,777\,216$. Pure red is $\{255,0,0\}$, pure green is $\{0,255,0\}$, and pure blue is $\{0,0,255\}$.

The 12 hues have one RGB value equal to 0, the second equal to 255, and the third value is 0, 255, or 127 (Figure 7.19). These pure colors are at the top of

FIGURE 7.18   Color wheel. (From Rezende (2013).)

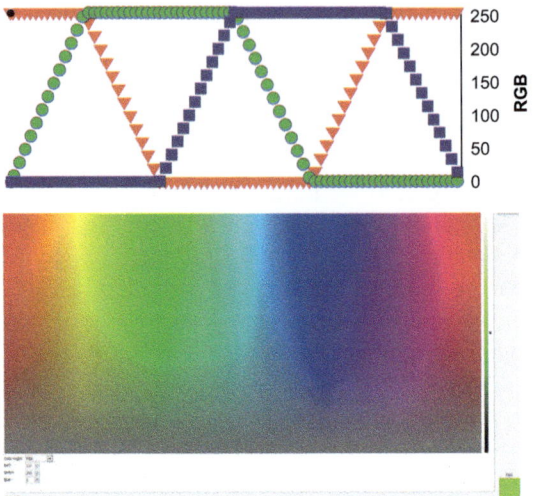

FIGURE 7.19 RGB color contour. The *x-y* plot represents the proportions of RGB from the very top of the contour plot. Triangles are red, circles are green, and squares are blue. As one progresses downward along the contour plot axes, colors become less vibrant: at the bottom the values of R, G, and B are equal, and the color is gray.

the contour plot. As you descend vertically, all of the RGB values tend toward 127—gray, which has an equal proportion of red, green, and blue—$\{x, x, x\}$. Shades of a graph begins from $x < 127$ and gray tints are from $x > 127$. Black and white are special cases of gray with RGB values $\{0, 0, 0\}$ and $\{255, 255, 255\}$, respectively, for additive colors. For subtractive colors, the CMY values are opposite: white is 0 and black is 255.

Color Palettes

Colors and *contrast* enhance the perception and cognition of an object (Dreyfuss, 1984). Objects in a presentation represent your ideas and work. *Simultaneous contrast* is how two near objects of different color affect each other (Chevreul, 1855).

Aesthetic color palettes are created on the basis of monochromatic, analogous, and complementary color schemes. On a color wheel, monochromatic colors rest along a straight trajectory from the center (white) toward the periphery (black) (Figure 7.18). In the hue, saturation, luminosity color scheme, monochromatic colors have the same hue and saturation, only the luminosity changes. Analogous color palettes consist of adjacent hues (the 12 colors in Figure 7.20). Generally, analogous color palettes are at the same circumference

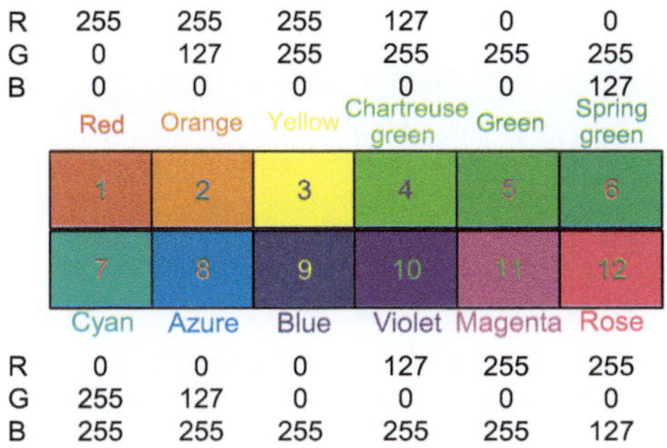

R	255	255	255	127	0	0
G	0	127	255	255	255	255
B	0	0	0	0	0	127
	Red	Orange	Yellow	Chartreuse green	Green	Spring green
	1	2	3	4	5	6
	7	8	9	10	11	12
	Cyan	Azure	Blue	Violet	Magenta	Rose
R	0	0	0	127	255	255
G	255	127	0	0	0	0
B	255	255	255	255	255	127

FIGURE 7.20 The colors in the bottom row are complementary to those in the top row. The color of the block number is the same as the complementary color of the block.

on the color wheel, which means they have the same saturation or chromatic level. Complementary colors are on the opposite sides of the color wheel. Combining complementary additive colors gives white, while combining complementary subtractive colors gives black.

Colors stimulate the mind, so choosing them is important. Black on white gives the highest contrast, but that will bore your audience. In contrast, too many colors will overstimulate the audience members, so they might not identify important information and eventually they will tune out. Excluding black and white, color palettes should have no more than four colors and preferably only three. Choose color palettes that ensure good contrast, which is necessary to enhance an object's visibility.

Monochromatic: A pure monochromatic color can be monotonous, but combining shades, tints, and saturation with black and white can be striking. *Monochromatic* color palettes consist of colors belonging to the same radial trajectory from the center of the color wheel (Figure 7.18). On the color contour plot, raising the scroll bar to the left increases the luminosity (tint), and lowering it decreases the luminosity (increases the shade) (Figure 7.19). Moving the cursor up is the equivalent of going toward the center of the color wheel. Moving it down is like going toward the edge.

The RGB values increase proportionately toward {255, 255, 255} with increasing tint, and decrease proportionately to {0, 0, 0} with increasing shades. The background in Figure 7.21 is purple—{127, 0, 255}. To create purple shades, green remains at zero and the blue value drops twice as fast as the red value. From the background {127, 0, 255} to block 8 {106, 0, 213}, blue dropped by 42, red by 21. From block 8 {106, 0, 213} to block 9 {85, 0, 171}, blue dropped by 42, red by 21. Blocks 7 to 12 are monochromatic purple shades.

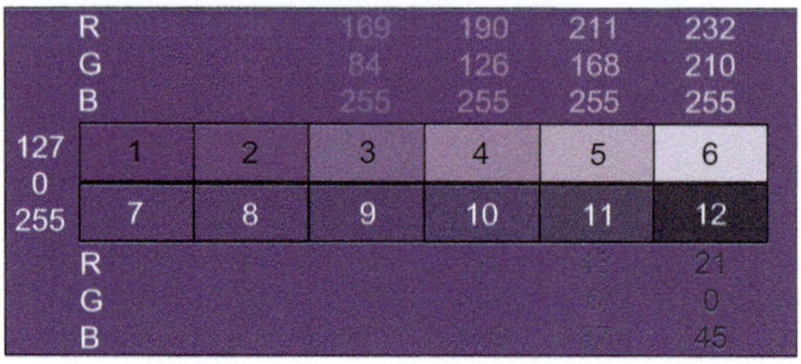

FIGURE 7.21 Background purple: {127, 0, 255} (RGB). To produce darker monochromatic purple, decrease the blue value twice as much as the red value and keep the green value at 0. At {0, 0, 0} the color is black. To produce lighter shades of purple, keep the blue value at 255 and increase the green value twice as much as the red value. At {255, 255, 255} the color is white.

To produce monochromatic purple tints, the blue value stays at 255 and for every increase in the red value, the green value is increased twice as much (blocks 1 to 6 in Figure 7.21).

Contrast: Contrast increases as the sum of the absolute values of the differences of the RGB values increases:

$$\Delta RGB_{i-j} = |R_i - R_j| + |G_i - G_j| + |B_i - B_j|.$$

Black and white have the highest contrast:

$$\Delta RGB_{white-black} = |255 - 0| + |255 - 0| + |255 - 0| = 765.$$

Contrast between the color in the top blocks and the black numbers (1, 2, 3, 4, 5, 6) is acceptable for all purple shades (Figure 7.21). Blocks 1 and 7 have the lowest contrast: $\Delta RGB_{1-black} = \Delta RGB_{white-7} = 413$. The RGB values of the block purple tints appear above the top row of blocks and the purple shades are below the bottom row of blocks. The RGB values of block 1 are illegible because the contrast is insufficient:

$$\Delta RGB_{1-bg} = \{137, 21, 255\} - \{127, 0, 250\} = 31.$$

The contrast between the block color and background is reasonable in block 4 ($\Delta RGB_{4-bg} = 189$). It is much better in block 5 ($\Delta RGB_{5-bg} = 252$). ΔRGB is the same for blocks 4 and 10, and blocks 5 and 11. As a rule of thumb, if $\Delta RGB > 250$, the contrast is acceptable, but higher contrast is better.

Complementary: Complementary colors lie on opposite sides of the color wheel (Figure 7.18). Mixing additive complementary colors produces white {255, 255, 255}, whereas mixing subtractive complementary colors produces black. Complementary colors have the maximum contrast. The RGB values of lime green are {127, 255, 0} and its complement is purple—{127, 0, 255} and $\Delta RGB = 510$. The colors at the bottom of Figure 7.20 are complementary to those in the top row. The pairs of complementary colors with primary colors have $\Delta RGB = 765$ and the others have $\Delta RGB = 510$.

R	239	16	120	8	247	135
G	15	240	8	120	133	247
B	116	139	59	69	185	196

1	2	3 (6)	4 (5)	5 (4)	6 (3)

| L, % | 100 | 100 | 50 | 50 | 150 | 150 |

FIGURE 7.22 Complementary colors with tints and shades. Blocks 1 and 2 are complementary colors with 100% luminosity. Block 3 is a shade of block 1 with 50% lower luminosity Block 5 is a tint of block 3 with luminosity 150% higher than that of block 1. Blocks 4 and 6 have the same luminosity and saturation as blocks 3 and 5 with a complementary hue. Block 6 is the complementary color of block 3, and block 5 is the complementary color of block 4.

Images and letters with $\Delta RGB > 500$ are uncomfortable to stare at for too long, so limit the number of slides with a high contrast to three or fewer. Too many slides with complementary colors will overstimulate the audience.

You can nuance the colors by adding gray (tones), or shading or tinting. Calculating complementary colors with shades or tints adds an additional parameter. The complement of fuchsia {239, 15, 116} is {16, 240, 139} at full saturation (100% luminosity) (blocks 1 and 2 in Figure 7.22). This fuchsia hue with half the luminosity becomes dark red and has an RGB value half that of the original: {120, 8, 59} (block 3). Block 4 has the complementary hue of block 3 but the same luminosity and saturation and is green. The numbers 3 and 4 in these two blocks have the complementary color of the other (the number 3 has the color of block 4 and the number 4 has the color of block 3). The contrast between blocks 3 and 4 has $\Delta RGB = 234$, which is below our minimum criterion of 250 and is difficult to see.

The RGB value for fuchsia with 50% more luminosity looks more like pink with an RGB value of {247, 133, 185} (block 5). Block 6 has the complementary hue to block 5 but with the same luminosity and saturation. The contrast for blocks 5 and 6 is better than that for blocks 3 and 4, but ΔRGB is the same.

The true complementary color for block 3 is the color in block 6, and for block 4, it is the color in block 5. The contrast between the block colors and the complementary colors (numbers in parentheses) is excellent, and $\Delta RGB = 391$.

Monochromatic and complementary: In Figure 7.23 the azure letters {0, 127, 255} are complementary to the orange background {255, 127, 0} and $\Delta RGB = 510$ (double our minimum recommended guideline). In the top row of blocks, the orange tint approaches white, and the orange shade in the bottom row approaches black. The change in the orange shade in the block is proportional to the change in the azure shade in the block numbers. The RGB values at the top refer to the numbers in the top row of blocks. The RGB values at the bottom refer to the numbers in the bottom row of

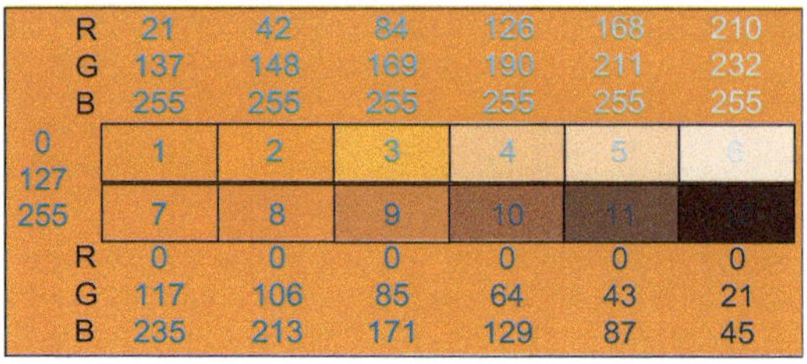

FIGURE 7.23 Background orange: {255, 127, 0} (RGB). For darker orange monochromatic colors, keep the red value at 0 and decrease the blue value twice as much as the green value. For lighter shades of orange, keep the blue value at 255 and increase the red value twice as much as the green value.

blocks. For example, number 3 is {84, 169, 255}, and the orange shade in block 3 is {255, 169, 84}; number 10 is {0, 64, 129}, and the orange shade in block 10 is {129, 64, 0}. The contrast of the azure and orange shades in the last two blocks is poor. Although the hues of the orange color in blocks 5, 6, 11, and 12 are complementary to the hue of azure, the contrast is low and thus the numbers are hard to read ($\Delta RGB_5 = 174$ and $\Delta RGB_6 = 90$). The contrast in block 4 (and block 10) meets the minimum criterion of 250 ($\Delta RGB_4 = 258$). Blocks 1, 2, and 3 have ΔRGB equal to 468, 426, and 342, respectively, and have great contrast.

Analogous: Analogous color palettes are common in nature, and are considered to be the most attractive. Adjacent colors on the color wheel are analogous. Three-color analogous color palettes include rose-red-orange, spring green-cyan-azure, and blue-violet-magenta (Figure 7.20). Choose one color that dominates the other two colors. The second color supports the dominant color, and the third color should be used sparingly for highlighting.

Split Complementary: Because split-complementary color palettes are hard to get wrong versus other schemes, researchers should concentrate on this palette first. Complementary palettes have two colors that are opposite on the color wheel with primary, secondary, and tertiary hues. The split complementary has three colors in which two are adjacent to the complementary color: red-spring green-azure, green-rose-violet, and blue-orange-chartreuse green are three examples.

7.4 DELIVERY

The first step to make a great presentation is to focus the message to suit your audience. Hettiarachchi (2014), who won the Toastmasters International World Championship of Public Speaking (2014), claims that a speech is a conversation with the audience rather than a monologue. Papers are also conversations with the reader. Unlike in papers, in presentations you repeat your message using a technique that comedians refer to as *callback*, where you allude to something you stated previously: *Remember that I mentioned....* When you conclude the talk, refer back to the summary slide. In fact, you can redisplay it and add some perspective, including uncertainties or opportunities.

Delivering the message eloquently is the second step of a great presentation. By eloquent, we mean naturally—practice being yourself in front of people. Speaking well is not an accident. You have to try, fail, repeat, and improve. Perhaps it won't take 10 000 h to master speaking in front of strangers (Gladwell, 2008), but even senior professors and academic staff seek better ways to communicate with students. Speak enthusiastically because your passion is contagious. The audience can't help but feel and share your excitement.

Practice making apparently boring subjects interesting to random people, colleagues, family, and friends. Make a video, identify your idiosyncrasies (*um, uh, right, you know*), and remove them (Day and Gastel, 2011). The first two sentences are the hardest to get out, but as soon as you begin talking, speaking becomes fluid. That is why you start with pleasantries, something easy to say: *I am happy to be here, The conference venue is fantastic, The previous speaker said...and I will continue this topic but....* It also establishes your conversational tone. Recognizing related work from the conference contextualizes what you are talking about, and immediately that shared experience connects you with your audience.

Often, a chairperson will introduce you, your institute, and the title of your presentation. Don't repeat what the chairperson has said, especially if you have less than 20 min.

One measure of enthusiasm is how fast you speak. Slow speakers are often dull and monotonic. If you are an insomniac, tape these talks and play them at night to help you fall asleep. You must vary your tone, speed, and volume.

Give your audience time to examine complicated graphs. If you have some text, walk the audience through it. Keep the audience's attention and don't allow it to wander. Just because you have said something doesn't mean your audience has understood it or agrees with you. Repeat what you want you audience to remember. Ask rhetorical questions to get people to focus on what you are talking about. The scientific community is skeptical and will analyze your data and identify trends independently of what you say. This takes time, and for this reason you have to go more slowly. The best way to control your pace is to say less than you plan to say. That way, you put less stress on yourself to get through all of the material.

We measured how fast people presenting at conferences spoke, and concluded that anything below 100 words per minute was kill-me slow and anything above 160 words per minute was too fast. When people spoke faster than this it was hard to measure accurately. People read at an average of 200 to 240 words per minute. Imagine trying to read a slide full of text, interpret graphs, and assimilate what a fast speaker says. Perhaps you can keep up for 2 min, or longer if you are very interested in the topic, but eventually you will tune out (or burn out).

If you speak fast naturally, slow down. If you can't, take more pauses. Hettiarachchi (2014) spoke at 120 words per minute during his award winning talk. The TED lecturers Al Gore, Hans Rosling, and Becky Blanton speak at rates between 130 and 160 words per minute, while other speakers exceed 180 words per minute (Dlugan, 2012). Steve Jobs spoke at 160 words per minute. People speak at rates above 160 wpm in casual conversations (Gallo, 2014) but audio books are best at lower rates (150 wpm) because people don't have the benefit of other sensory inputs like watching the lips, body language, facial expressions. TED talks average over 160 wpm but you have to slow down for scientific presentations. English is often a second or third language for many attendees at international conferences. An audience in Japan criticized a presenter who spoke English too quickly. The audience felt that the speaker disrespected them.

In plenary lectures, speakers have time to dwell on the history of the field and current advances. They must also outline emerging and challenging areas that remain to be studied. Some eminent researchers advertise their contributions in plenary lectures and keynote speeches, like a soapbox in Hyde Park. Offer a broad view of what is, but talk about what is left to do.

A senior manager from General Motors gave a plenary lecture at a conference in Niagara Falls in 2014. He spoke at a rate exceeding 160 words per minute for 1 h with almost 100 slides. The presentation was excellent: the slides were engaging, colorful, and clear; what he said was accessible; the scientific value was tremendous. He cited many recent publications in top journals. However, the volume of information was overwhelming. Speaking at 160 words per minute for 1 h means you will say 9600 words, which is equivalent to three scientific papers.

Tips

1. Thank the chairperson for the introduction and the opportunity to speak at the conference.
2. Address the audience members (choose from among the following):
   - Thank them for coming.
   - Tell them you are pleased to be there.
   - Mention something about the conference or the chairperson.
   - Give a personal anecdote or comment.
   - Recognize speakers preceding you (rephrase what they have said that is related to your work).
3. Speak enthusiastically (don't read the screen or notes).
4. Speak no faster than 150 words per minute and no more slowly than 120 words per minute.

5. Project your voice with confidence (even with a microphone).
6. Pause for 1–2 s every minute or two (pregnant silence).
7. Vary the volume.
8. Vary the pace.
9. Ask rhetorical questions (and pause).
10. Express the concepts with plain language and avoid jargon.
11. Say less than you plan to say.
12. Practice and get feedback.

Questions and Answers

After you have finished your brilliant exposé, the next ordeal is to respond to questions. You are not as smart as the collective intelligence of the group of scholars who have gathered to listen. People have questions, but few are motivated to ask them. Most times someone will ask something, but sometimes no one will. It is a bad sign if no one asks a question. Here are reasons why no one may have asked a question:

1. No one cared what you were saying (poor-quality slides).
2. No one believed what you were saying (poor slides).
3. No one understand a word of what you said (poor delivery).
4. You looked too frail to accept a question (not enough preparation).
5. There was not enough time (you spoke too slowly or you had too much information).
6. People are scared of you (too aggressive).
7. The organizers put your talk with unrelated talks.
8. Your talk was brilliant and it left everyone speechless (you must have read this chapter).

Questions may be general comments about the presentation or the results, or they may be about something else the audience found remarkable. As a speaker, all questions are good; at least someone was listening.

People ask questions for the following reasons:

1. You were unclear.
2. The details on the method or the results were insufficient.
3. You made errors.
4. They doubt what you say.
5. They disagree with what you say.
6. They seek your opinion on related topics.
7. They are interested in more detail.
8. They require more contextual information.
9. They are curious.
10. They want to talk about their own work (grandstand).
11. They want to congratulate you on a brilliant speech.

Although question and answer sessions can be adversarial, empathize with people asking questions. Wait for them to finish their sentence before you respond—do not interrupt, even to help them if they are struggling for a word (Davis, 1997). Ask the person to repeat it if you don't understand. If you still don't understand, ask the chairperson to help (Silyn-Roberts, 2013). To respond to a question, first thank the person and even repeat it so everyone hears it clearly. Be polite: people are less likely to remember your response if you are impolite. Conferences are full of people who might like to collaborate with you, hire you, or cite your work. So, don't be aggressive, and take as much care in how you respond as in what you say.

It doesn't have to be embarrassing if you don't know the answer. Choose among the following responses:

1. I don't know (a good response but tough to say).
2. Interesting question, let me think about it (this also means I don't know).
3. I don't have an answer for that, can we talk after the session.

Alternatively, you could think on the spot and develop a brilliant response. This option is dubious. After you have been speaking and standing, your brain is not functioning at 100%. Stick to one of the first three responses when you don't know the answer.

Posture and Gestures

Choose several individuals in different places in the hall and talk to them as if they were friends. Shift your focus from one person to another occasionally. Stay away from the podium, get close to your audience, and move as you talk. Don't exaggerate like Cussler (a professor from the University of Minnesota), who would throw himself to his knees.

Your arms shouldn't be plastered to your sides like a mummy. Reinforce what you say with hand and finger gestures. Go to one side of the screen, and when you present a graph, state the title, point to the axes, and say what they represent and their minimum and maximum values. For example, "The x-axis represents years from 1950 to 2010 and the y-axis represents the progression of computational power in teraflops…." Avoid talking to the screen or keeping your back to the audience.

Smiling and enthusiasm are contagious. If you look like you are enjoying what you are talking about, the audience can't help but feel enthusiastic too. We would rather spend time with people who make us feel good and have a positive outlook.

Remember that people's attention span is very short. On electroencephalograms, the high-frequency gamma waves, associated with perception, consciousness, and higher mental activity, drop significantly when one is watching television. Researchers connected a device that measures brain activity to children's heads. The researchers turned on a television and told the children that it would shut off if they stopped concentrating. The children were able to maintain the television on for a mere 24 s. Lady Gaga in the video of "Applause" changes the image every second. Even though our attention span is short, that is too fast; change slides at a frequency between 1 per min and 2 per min.

Respect the time and request that the chairperson give you a sign when 5 min remain for the talk and also at 1 min from the end. If the presentation goes beyond the time limit, this is disrespectful to the next speakers and to the audience. Many conferences have parallel sessions, and delegates select presentations in different sessions. Going overtime makes this difficult to do and is disruptive. Any distraction is annoying to the speaker and the audience.

7.5 EXERCISES

1. *Introduction*: Improve the slides in the following figures:
   a. Figure 7.24,
   b. Figure 7.25.
2. *Methods/Experimental*: Identify positive and negative aspects in the following figures:
   a. Figure 7.26,
   b. Figure 7.27.
3. *Results*: What is wrong with the graphs in the following figures:
   a. Figure 7.28,
   b. Figure 7.29,
   c. Figure 7.30.

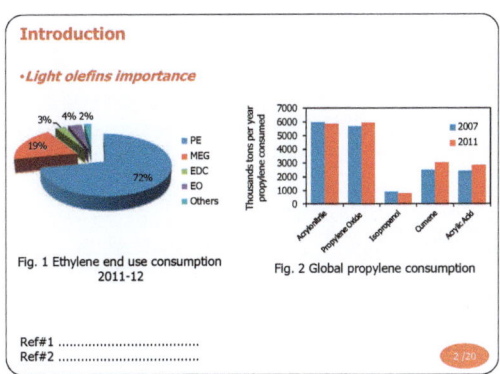

FIGURE 7.24   Importance of light olefins.

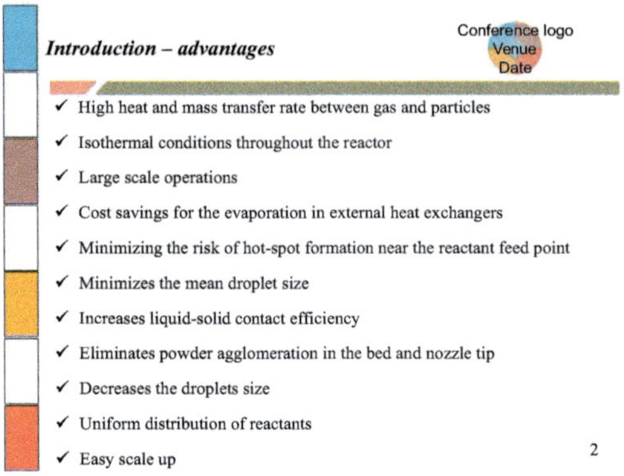

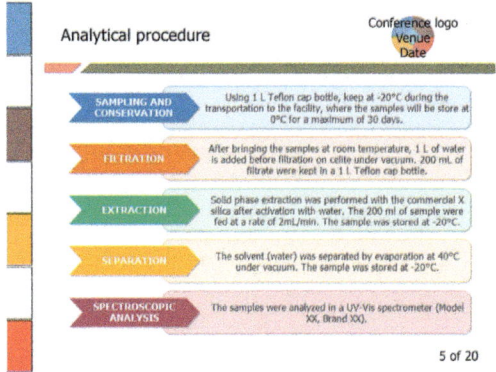

Introduction – advantages
Conference logo
Venue
Date
✓ High heat and mass transfer rate between gas and particles
✓ Isothermal conditions throughout the reactor
✓ Large scale operations
✓ Cost savings for the evaporation in external heat exchangers
✓ Minimizing the risk of hot-spot formation near the reactant feed point
✓ Minimizes the mean droplet size
✓ Increases liquid-solid contact efficiency
✓ Eliminates powder agglomeration in the bed and nozzle tip
✓ Decreases the droplets size
✓ Uniform distribution of reactants
✓ Easy scale up
2

FIGURE 7.25 Introduction—advantages.

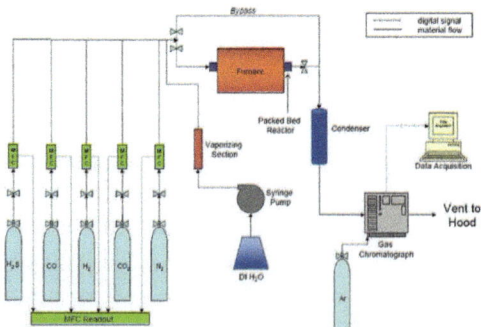

Analytical procedure
Conference logo
Venue
Date
SAMPLING AND CONSERVATION
Using 1 L Teflon cap bottle, keep at -20°C during the transportation to the facility, where the samples will be store at 0°C for a maximum of 30 days.
FILTRATION
After bringing the samples at room temperature, 1 L of water is added before filtration on celite under vacuum. 200 mL of filtrate were kept in a 1 L Teflon cap bottle.
EXTRACTION
Solid phase extraction was performed with the commercial X silica after activation with water. The 200 ml of sample were fed at a rate of 2mL/min. The sample was stored at -20°C.
SEPARATION
The solvent (water) was separated by evaporation at 40°C under vacuum. The sample was stored at -20°C.
SPECTROSCOPIC ANALYSIS
The samples were analyzed in a UV-Vis spectrometer (Model XX, Brand XX).
5 of 20

FIGURE 7.26 Analytical procedure.

Bypass
digital signal
material flow
Furnace
Packed Bed Reactor
Condenser
Vaporizing Section
Syringe Pump
DI H₂O
Data Acquisition
Vent to Hood
Gas Chromatograph
CO
Ar
MFC Readout

FIGURE 7.27 Experimental reactor schematic including gas manifold and analytical instrumentation.

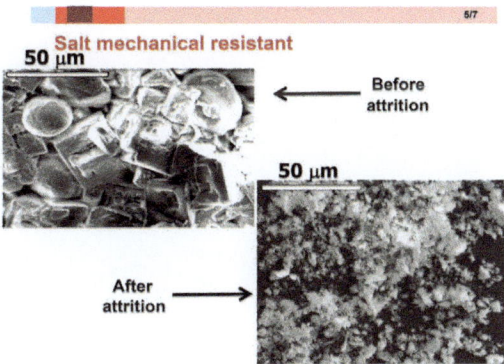

FIGURE 7.28 Catalyst and salt attrition.

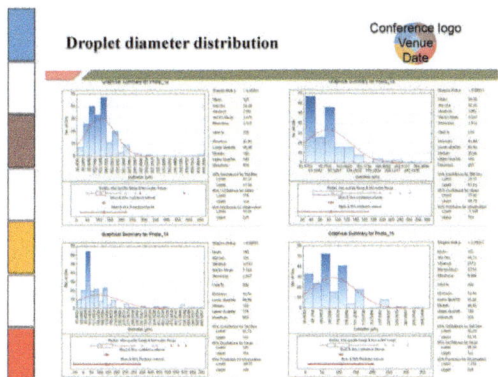

FIGURE 7.29 Droplet diameter distribution.

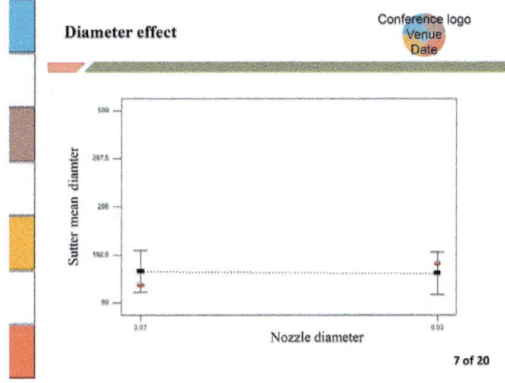

FIGURE 7.30 Diameter effect

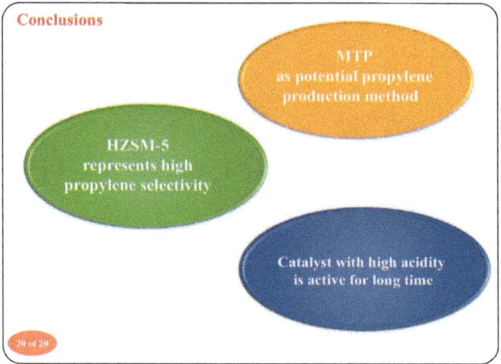

FIGURE 7.31   Conclusions.

4. *Conclusions*: How can you improve these conclusions?
   a. Figure 7.31.
   b. Future research can be made in three directions:

   - Improving the existing model;
   - Expanding the model to a three dimensional model; and
   - Building a mechanical model.

5. *Colors*:
   a. What is the color palette of Figure 7.5? Comment on the choice of color.
   b. Which color is complementary to the blue background in Figure 7.9?
   c. How would you improve Figure 7.11, including the colors?
   d. Create a monochromatic color scheme with fuchsia.
   e. Make a complementary color scheme for green {0,255,0}.
   f. Propose a three-color analogous palette with azure.
6. Train yourself to make the most boring talks interesting. Read a newspaper article and explain it to friends.
7. Choose a 15 min presentation, reduce it to 5 min and five slides, and present it to your parents, family, or friends. Ask them what they did not understand.
8. Not everyone will be enthusiastic about your work. Prepare yourself for probing (aggressive) questions. Respond to the following:
   a. This work is elementary and is covered in a fourth year undergraduate course.
   b. Is that true?
   c. Work by others in this area shows the exact opposite trend.
   d. Why have you neglected the contributions of…?
   e. What are the experimental errors?
   f. What is the uncertainty in the experimental data?
   g. How can you be sure that the analytical techniques are adequate?
   h. Your results are unimpressive versus what…demonstrated.

i. How can you extrapolate your results to a larger scale?

j. You have presented interesting experimental data. Have you derived a mechanism that accounts for the trends in the data?

k. Why have you ignored the factor (temperature, pressure, viscosity, etc.) in the experimental plan?

l. How did you validate your assumptions?

Posters That Captivate

An oral presentation is more prestigious than a poster presentation, especially in conferences with fewer than four parallel sessions. Most would rather present their work in front of a large audience and consider a poster as the consolation prize.

Many industry researchers prefer posters because they promote one-on-one interactions. Moreover, they have more time to get approval from management and legal counsel. If management changes its mind at the last minute, withdrawing an oral presentation disrupts the conference program, whereas the consequences of canceling a poster are less severe. Many conference organizers are changing their format, and include oral presentations with posters to ensure everybody has time to ask questions and interact with the speakers.

For some conferences, the technical committee requires full manuscripts and/or the extended abstracts as much as 6 months in advance. Reviewers rate the manuscripts' originality, impact, scientific merit, and clarity. An advisory board ranks the manuscripts on the basis of the reviews, but also considers authors' reputation, the number of submissions from the same research group, and geographic equality when preparing the technical program. The review process ensures that the best papers reach the widest audience. Furthermore, publishing the program in advance with the top names attracts people who might not have otherwise attended. Members of organizing committees and scientific advisory boards have a high success rate in securing oral presentations. A blind review process would be more impartial, thus ensuring a higher-quality conference.

Communicate Science Papers, Presentations, and Posters Effectively. http://dx.doi.org/10.1016/B978-0-12-801500-1.00008-5

Weighting a research group's reputation, which is based on a track record of contributions, makes sense. Often these groups come with large delegations, which helps finance the conference. However, established conferences can stagnate if organizers favor well-known individuals excessively. Dynamic young researchers will seek alternative venues to communicate their work.

An excellent extended abstract demonstrates novelty, scientific impact, and clarity. Scientific committees reject work that doesn't satisfy these criteria. If you think that reviewers or the committee has underrated your document, contact the conference chair. Committees review hundreds of papers and are likely to overlook some outstanding research. Highlight the merit of your work and insist that it is a valuable contribution to the conference. Conference chairs have flexibility in accepting papers for various reasons. They are interested in welcoming as many people as possible.

Presenting a poster is an opportunity to talk with conference attendees one-on-one. Professors and academic staff look for graduate students and postdoctoral fellows, and companies look for employees. You can engage people and discuss your work in a nonthreatening context. Be ready to argue, accept criticism, and teach. Poster sessions can be dynamic, and you might not always be in front of your poster. That's why posters must be self-explanatory. Some people prefer to look at posters uninterrupted, and most will just glance at them: posters must be engaging enough to attract attention and to convey the essential theme.

Why will anyone stop in front of your poster to appreciate your contribution to science? We surveyed 65 engineering and chemistry graduate students, postdoctoral fellows, and professors to find out (Table 8.1). They are likely to come if they work in the same area or are interested in the topic. They are unlikely to come just because many people are looking at your poster, or because you are from a well-known institute. They avoid posters that have dense text, but are attracted to posters that are aesthetic, clear, and logically organized. But they might not come, even if they are interested in your research, if your poster doesn't look like one.

8.1 CONTENT

People often spend less than 90 s in front of posters while they walk around the hall (O'Connor, 1991). Keep this in mind while designing a poster. How many themes, ideas, and concepts and how much experimental data can one possibly assimilate in 90 s? Three key results—important contributions or insights—are sufficient to engage your audience; people who look at your poster might even remember them all. They cannot read 300 words in 90 s; better to keep the word count below 150.

Since all the information is on one page, connect graphs, tables, and texts that are related with arrows. This construct is difficult to implement in presentations, but is a powerful tool in posters.

TABLE 8.1 Engineers and Chemists Told Us Why They Stop in Front of Posters—Mostly It Is Because the Subject Is Similar to Their Own Work or They Like the Topic (N = 65)

You Stop in Front of a Poster Because	Strongly Agree	Agree	Ambivalent	Disagree	Strongly Disagree
The subject is similar to your research subject	34	19	5	3	2
You are generally interested in the topic	19	40	3	0	3
You are familiar with the subject	19	29	11	4	2
The poster is aesthetically appealing	17	16	11	9	2
You know the presenter	16	20	13	9	1
You want to learn something new	14	23	16	7	4
The images are clear, large, and colorful	12	24	14	9	4
The poster is logically organized	8	26	13	13	3
The presenter is good-looking	5	17	12	13	13
It is from a renowned institution	3	19	21	17	3
Many people are looking at it	2	16	21	16	8
The text on the poster is dense	1	3	6	17	19

Banners

Poster banners include titles, authors, affiliations, and logos, and at times, conference details. Titles for posters, like those of papers, are longer than for presentations (Chapter 6). In presentations, speakers abbreviate titles so that the audience pays attention to what they say and not to what they wrote. Concentrate on keywords; if the title has more than 16 words, you have to shrink the text to fit it into the banner, which will make it difficult to read from 3 m away. When you present posters at major international conferences with thousands of people (e.g., American Chemical Society conferences), make the title accessible to as many people as possible. Most people won't be familiar with the specifics of your subject, but you still want them to stop to look. If they don't understand your title they probably won't stop to look (unless they fall into one of the categories in Table 8.1). At specialized conferences with fewer than 200 participants, esoteric titles are fine—but eliminate all zero words such as *results*, *toward*, and *study*. This title is from the International Conference on Topics in Astroparticle and Underground Physics, and contains zero words:

— First results in Pure High Pressure Xe towards the study of Xe + TMA mixtures with Dark Matter directionality sensitivity and supra-intrinsic energy resolution for $0\nu\beta\beta$ decay searches (Oliveira et al., 2013).

For general conferences, the title should be simpler and adapted to an extended audience. Regardless of the conference, this 26-word title is too long. Constrain the title to two lines of text. The title font size is over 72 pt, and thus you waste space with multiple lines of text—the information density of a large banner is lower than that of the body of the poster.

Section Titles

Poster content is the same as in articles and presentations, but the form goes beyond that of either. The content includes an introduction, experimental methods, results, conclusions, references, and acknowledgments. There is no abstract: the poster itself resembles an abstract because it contains the same elements, but in a visual form. Like in presentations, replace the boilerplate section titles with text pertinent to the paper. For instance, replace "Experimental" with the name of the principal equipment, instrument, or procedure—"Thermogravimetric curves," "Antioxidant activity," or "Sol-gel synthesis." Here is an example of section titles for a paper on developing new textile fibers:

IMRAD title	Alternative title
Introduction	Fiber Chemistry
Methods (Experimental)	Physical Properties
Results	Performance Criteria
Conclusions	Swimwear End-use

Each of these section titles correlates with the IMRAD nomenclature, but they are expressed creatively. Neither *Science* nor *Nature* signposts its articles with IMRAD titles. Precise titles orient your viewers without wasting space.

Context (Introduction)

Papers dedicate several pages to introductory comments, background, literature review, and challenges—these sections frame the paper in a larger context. Posters are constrained by space and time. You cannot copy-paste the extended abstract you submitted to the technical committee to serve as the introduction. Bullet points are better than full sentences, and images are better than bullet points. With three to six bullet points, you must do what a paper does in several pages. Emulate the abstract format of *Science*: in five sentences, summarize the entire paper, where each sentence represents one of the IMRAD sections. (*Nature* does it in eight sentences).

Figure 8.1 uses images and bullet points to introduce the subject (xylose). Posters present all the information at once, so viewers' attention can wander from one panel or image to another and can frequently return to the starting point. So, contrary to a presentation, don't repeat information.

Do not explicitly state hypotheses, purposes, and objectives. Leave these to high school science fairs.

- *The objectives were to determine…*
- *Our objectives were to demonstrate…*
- *We showed that…*

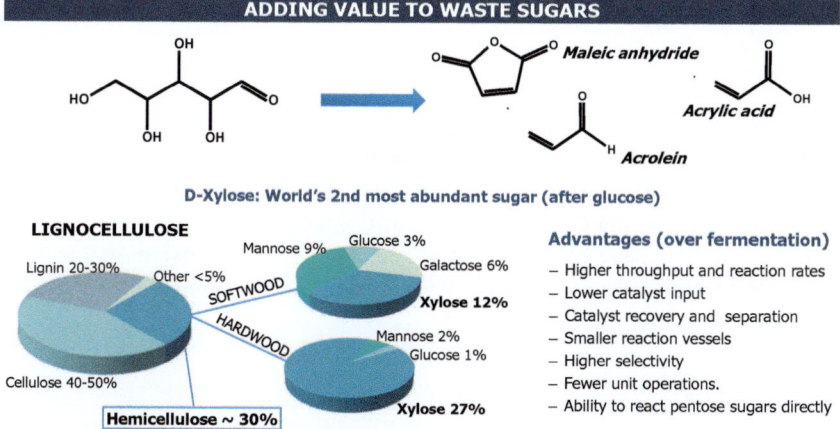

FIGURE 8.1 This visual introduction reduced the amount of text by a factor of 3. Xylose is derived from lignocellulose and is a potential feedstock to produce chemicals. (From Ghaznavi et al. (2014).)

TABLE 8.2 The Poster Is Entitled "Improving Impact Performance of Portable Electronic Devices"

Bullet Point	
1	Market is booming with 15% growth per year
2	PC/ABS and PA/ABS polymers have poor drop impact resistance
3	Polymer blends with Spandex fibers improve resistance by 40%
4	Smartphones and tablets most promising markets

Note: With these bullet points, explicitly stating a hypothesis or objectives is unnecessary.

These statements are metadiscourse. Let the data, the discovery, and the analysis speak for themselves (Table 8.2). The context section of a poster is equivalent to the Highlights section of a paper.

Methods/Experimental

Don't exceed 30% of the poster's space with the introduction and experimental sections. Leave the minutiae to papers.

Bullet points are acceptable in the introduction, but images must dominate the experimental section. Include schematic diagrams, pictures of instruments, and flow diagrams (and tables). Brand names and models are superfluous. Don't state that the experimental procedure is described elsewhere.

The methods section of a poster on exercise rehabilitation (Crouse et al., 2006) has 300 words—twice as many words as the total number of words we recommend for a poster. The entire poster has over 1300 words, which would require 7 min to read it if it were a novel or a newspaper. We summarized the methods section with six images (and titles) and five bullet points (Figure 8.2) in the same 45 cm × 50 cm frame.

Results

Focus the poster on the results: dedicate as much as 60% to 70% of the area of the poster to them. Graphs, charts, pictures, and tables must monopolize the space. Describe them with bullet points, not sentences. Remember, three concepts in a poster are enough. Don't crowd your poster with too many graphs or tables. If you have space, compare the data with the existing literature, highlight differences, and identify trends. Express the meaning of error bars and state the limitations (see Chapter 6).

Discussion/Conclusions

The discussion can go with the conclusions, with the results, or it can stand alone. What is important is that everyone understands what you have accomplished. Don't leave them guessing; state it explicitly.

Cardiac rehabilitation (CR)

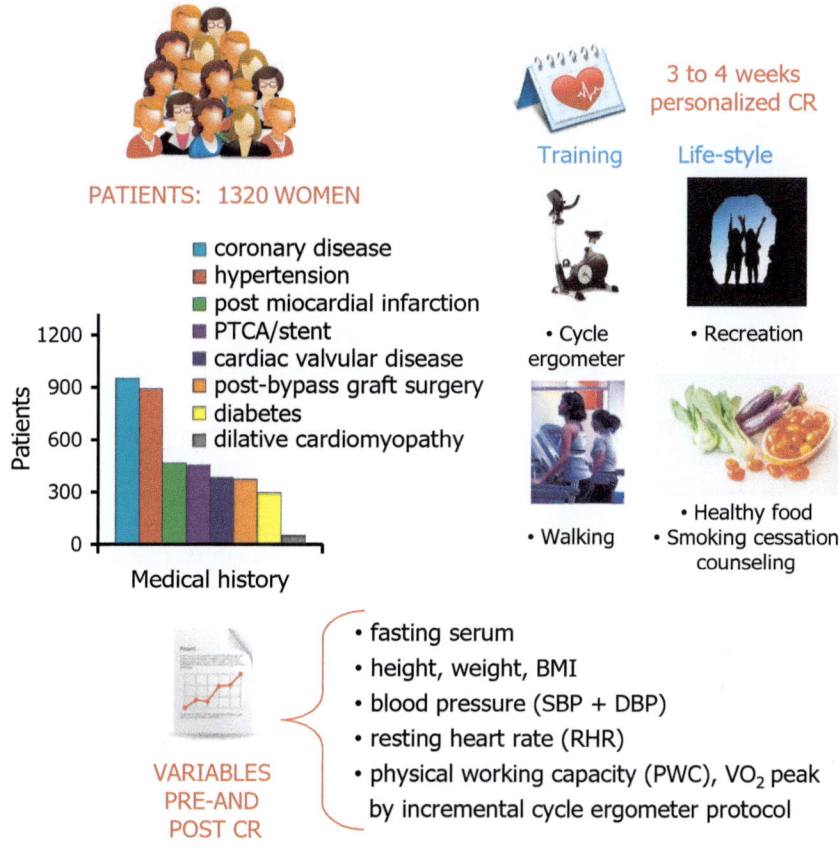

PATIENTS: 1320 WOMEN

- coronary disease
- hypertension
- post miocardial infarction
- PTCA/stent
- cardiac valvular disease
- post-bypass graft surgery
- diabetes
- dilative cardiomyopathy

3 to 4 weeks
personalized CR

Training   Life-style

- Cycle
ergometer   - Recreation

- Healthy food
- Walking   - Smoking cessation
counseling

Patients / 1200 / 900 / 600 / 300 / 0

Medical history

VARIABLES
PRE-AND
POST CR

- fasting serum
- height, weight, BMI
- blood pressure (SBP + DBP)
- resting heart rate (RHR)
- physical working capacity (PWC), VO_2 peak
  by incremental cycle ergometer protocol

Statistics: ANOVA, alpha = 0.05

FIGURE 8.2   Visual alternative to the 300-word methods section. (From Crouse et al. (2006).)

The discussion and the conclusions highlight the novelty of the work, its contribution to the field, and its economic incentive. All of the data should be in the results section, so don't repeat what you have already said. If the link between the results and the discussion/conclusions is ambiguous, connect lines from this section to images and graphs in the results.

Use the active voice—express the agent and the patient—and avoid metadiscourse (see Chapter 2).

The following sentence disregards many of our guidelines on writing: "On the basis of the experimental results obtained so far, it is possible to preliminarily conclude…."

The conclusions are obviously based on the experimental results. This sentence is hedging (*feeble*), and it wastes space (*redundant*). Clearly, the conclusions will not be based on results that the authors have not yet obtained.

The conclusions section is the only section that includes forward-looking statements. Choose a better label: consider for instance "Perspectives."

The poster of Patrick et al. (2013) on herbal supplements has 800 words and looks like the authors copy-pasted the text directly from a paper. In the conclusions, they repeat text from other parts of the poster in over 200 words and 11 sentences.

1. *High performance mass spectrometry has been used to differentiate various forms of common nutriceuticals with respect to their "nutrient" content.*

   This sentence repeats the title. (**redundant**, **passive**, prissy quotations).

2. *Differences in analyte concentration were measured by relative abundance of select analytes in the different....*

   This sentence repeats what is in the experimental section and is obvious. (**passive**)

3. *Identification was achieved using accurate mass determination with measured mass accuracies of typically less than 1 ppm.*

   The results section already contains this information. (**feeble**, **passive**)

4. *In addition, relative isotope abundance measurements provided a mechanism by which to support the formula of putative analytes.*

   Isotope abundance does not provide a mechanism. It is evidence that supports a mechanism. (**feeble**)

5. *The analysis also detected five analytes with well-defined m/z values but remain unknowns.*

   They could express this idea as follows: The UHPLC instrument detected five unknown analytes $m/z = \ldots$. Analysis doesn't detect analytes, the instrument does. (**feeble**)

6. *The analysis of Green Tea samples was achieved with both HPLC (not shown here) and UHPLC.*

   Conference participants come to poster sessions to see what's on posters, not what isn't. You must convey your message in the limited space of a poster. Do not waste it by stating what you are not going to report or talk about.

7. *The analysis time in UHPLC was less than 2 min at 40 spectra/second and provided the differentiating results shown above.*

   The experimental and results sections already contains this information. (**passive**, **redundant**)

8. *High performance mass spectrometry cannot provide direct evaluation of the nutritional content of natural extracts, but it is capable of providing a high information content fingerprint.*

9. *This fingerprint may ultimately provide the needed information on the key "active ingredients" in certain nutriceuticals.*

   These two sentences mention "provide" three times. Mass spectrometry cannot "evaluate"; rather, it measures.

10. *Theamine, theophylline, and caffeine were elevated relative to the other Green Teas in the organic green tea and in the Green Tea Supplement samples.*

    What are the implications?

11. *High performance MS was implemented using FFP^{TM} technology and provided acquisition speed, mass accuracy, and isotopic integrity with No Compromises.*

    This sentence is meaningful (but **passive**).

Here are the same conclusions, with three bullet points and 44 words:

- UHPLC analyses the complete spectra in less than 2 min at 40 spectra per second.
- FFP^{TM} technology provides acquisition speed, mass accuracy, and isotopic integrity with no compromises.
- High-performance MS provides fingerprints to identify active ingredients in green tea, ginger, and acai berry.

However, these conclusions are missing both the outlook and perspectives, and how the findings would improve the field of application.

Acknowledgments

Acknowledge organizations that have funded the work as well as the different institutions involved. Displaying the logo of these institutions on the slides is appropriate (see Chapter 6).

8.2 STYLE—FORMAT

We asked the participants in our survey why they stop in front of posters and what makes a poster engaging (Table 8.3). Almost all of them said that posters must be aesthetic and that poor-quality images are intolerable. Surprisingly, most agreed that objectives must be stated and that conclusions should be explicit. Fewer were so equivocal with respect to bullet points and whether or not the poster should be self-explanatory. Still, 80% agreed that bullets points are better than sentences and that they should be self-explanatory. Finally, just over 50% preferred the classic IMRAD format.

Posters are not abridged articles. A poster is a visual tool, so use it that way. Keep it simple, and remember that just because you can read it on your monitor does not mean it is legible in a conference room. Even though almost everyone agrees as to what makes a great poster, not everyone makes great posters.

TABLE 8.3 Engineers and Chemists ($n = 65$) Rated the Importance of Poster Content and Features

Poster Content	Strongly Agree	Agree	Ambivalent	Disagree	Strongly Disagree
Posters should be attractive	46	15	2	1	1
Poor-quality images are intolerable	45	16	0	1	3
Objectives must be stated	39	15	7	1	1
Conclusions must be explicit	36	19	4	3	2
Figures and tables should have captions	31	22	5	3	3
Bullet points are better than sentences	29	22	10	1	2
Posters must be self-explanatory	27	27	10	0	0
Headings state "Abstract," "Introduction," "Experimental," "Discussion"	9	28	13	9	6
Poster features	Ambivalent				
The font is large enough to read at __ m	7	33	14	2	2
More than __ tables is unacceptable	5	20	24	8	2
No more than __ concepts are desirable	3	18	29	5	3
Fewer than __ figures is insufficient	8	13	20	13	4

What makes a poster bad? Too much text, illegible text, too many images (too few images), columns of data, too many colors (not enough colors), poor contrast between background and text, too much detail, not enough context.

Most of the respondents of the survey thought that the text should be legible at a distance of 2 m to 3 m. The three structural features of posters are the number of tables, concepts and figures: over three-quarters of the respondents thought that more than three concepts or tables is too many. They favored figures, and 80% thought posters should have at least three figures. Images must dominate a poster. All posters should have at least five images: an image for the context, a couple of images for the experimental details, and some more images for the results.

From this survey, the ideal features of a poster are as follows:

- text legible at 3 m,
- three new concepts or fewer,
- three tables or fewer, and
- three figures or more.

Consider that a poster is like a presentation for which you have 5 min to convey your message. Five minutes means no more than five slides, and by association five panels. To consolidate our recommendation to favor figures and tables, substitute figures and tables for text in the introduction, methods section, and conclusions section.

Panels

The banner spans the top of the poster. Try to generate a logical sequence where the elements of the poster merge from one panel to another seamlessly. The standard Cartesian format reads left to right and top to bottom (Figure 8.3).

Seamless does not mean subdividing the poster into mundane subsections such as "Introduction," "Materials and Methods," "Results and Discussion," and "Conclusions." There are many effective formats for posters. Layouts are available online (Colin Purrington, 2014). Focus on three principal themes or contributions. White space is preferable to overcharging the poster with text and images, but distribute the content equally over the poster.

To verify that the poster is visually appealing, remove all the text (except the text included in the graphic elements—photos, charts, and figures). According to Colin Purrington, ex-professor of evolutionary biology at Swarthmore College, "If you removed all the info besides the graphics, the poster should still be pretty good" (Zielinska, 2011). Figure 8.3 shows a bad poster: take away the text and there is nothing left. Figure 8.4 has plenty of white space and images. The context is well laid out at the top, and the poster emphasizes the experimental method with two images in the center. The images are illustrative rather than quantitative. The authors report quantitative (comparative) data in a table. The table has too many lines and colors and too much text. Conclusions are missing,

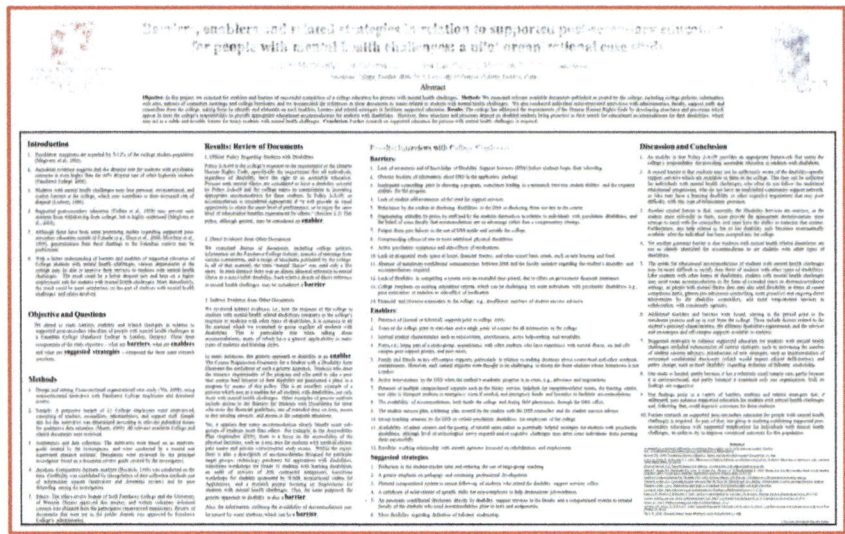

FIGURE 8.3 This is *not* a poster, rather it is an article printed on a big piece of paper with larger text.

so the viewer has to either listen to an author present the work or deduce the contribution from the data table and images.

Graphical abstracts can inspire poster formats, but remember that the orientation of graphical abstracts is landscape (Chapter 6). Figure 6.6 is an example of adapting a flowchart format to a poster (Valadés-Pelayo et al., 2015). At the right, arrows emerge from two blocks, and combine into a third block in the center. Another two arrows from the center block proceed to blocks 4 and 5 near the bottom and at the far right. The images tell the story, and the arrows lead the viewers through the process. Text is sparse, which is necessary for a graphical abstract, but more text would contextualize the subject.

Independent panels separate subject matter (Figure 8.5). In the top panel, Pirola et al. (2013) asks a question and then demonstrates the context of the problem with figures, arrows, and text. In the lower left panel, with two images, a table, and a schematic, they demonstrate how they approached the problem. Finally, in the bottom left panel, they summarize the results and conclusions with a table, a graph, and text. This is an excellent poster in that it minimizes text and has the ideal number of graphs and tables. It still might be too busy and it lacks white space.

Figure 8.6 has three panels. Swietoslawski et al. (2014) combine the synthesis (experimental) and results information in a single large panel below the title. The "Conclusions" and "Acknowledgment" panels occupy less space in two panels at the bottom. Their innovation is to use arrows to relate pieces of equipment at the bottom of the top panel. Above each image, they place

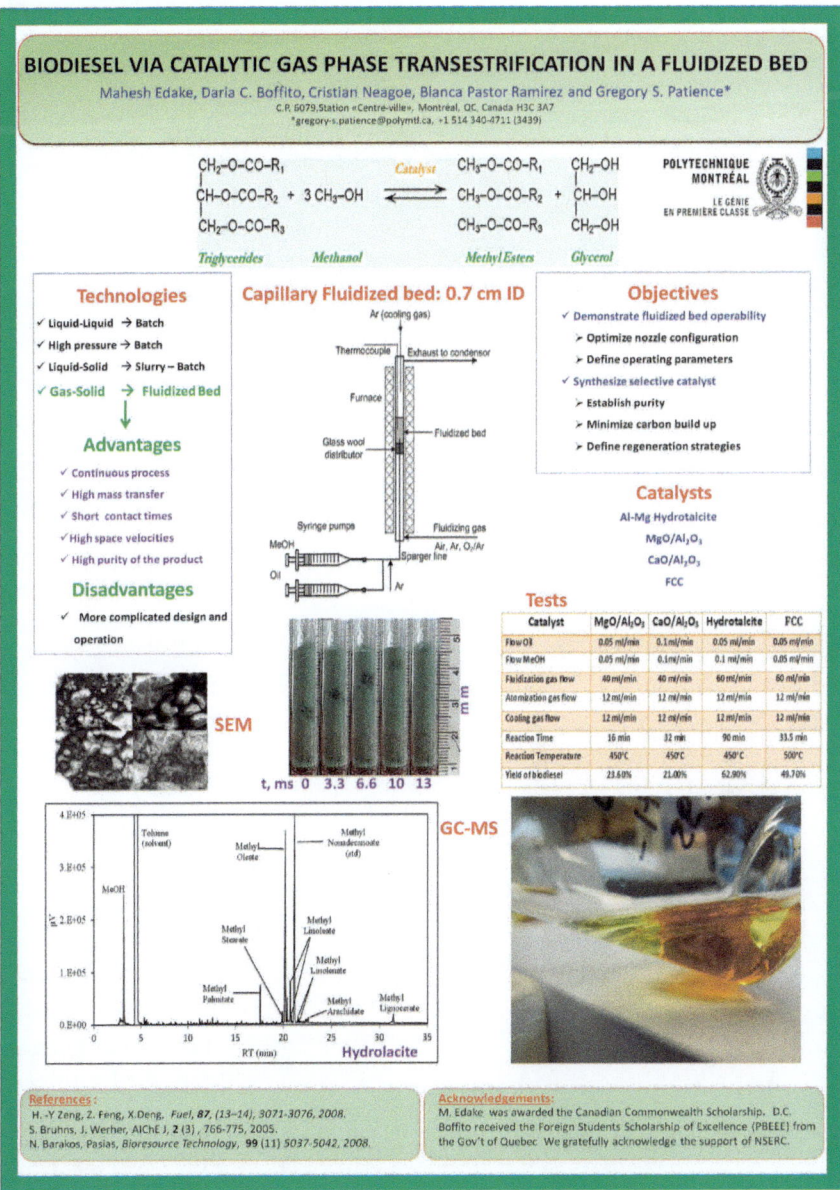

FIGURE 8.4 This poster has a minimum amount of text and many images. The poster focuses on microreactors, and two images fill the center. The table is poor, there are no *x-y* plots, and the conclusions are "missing." (From Edake et al. (2013).)

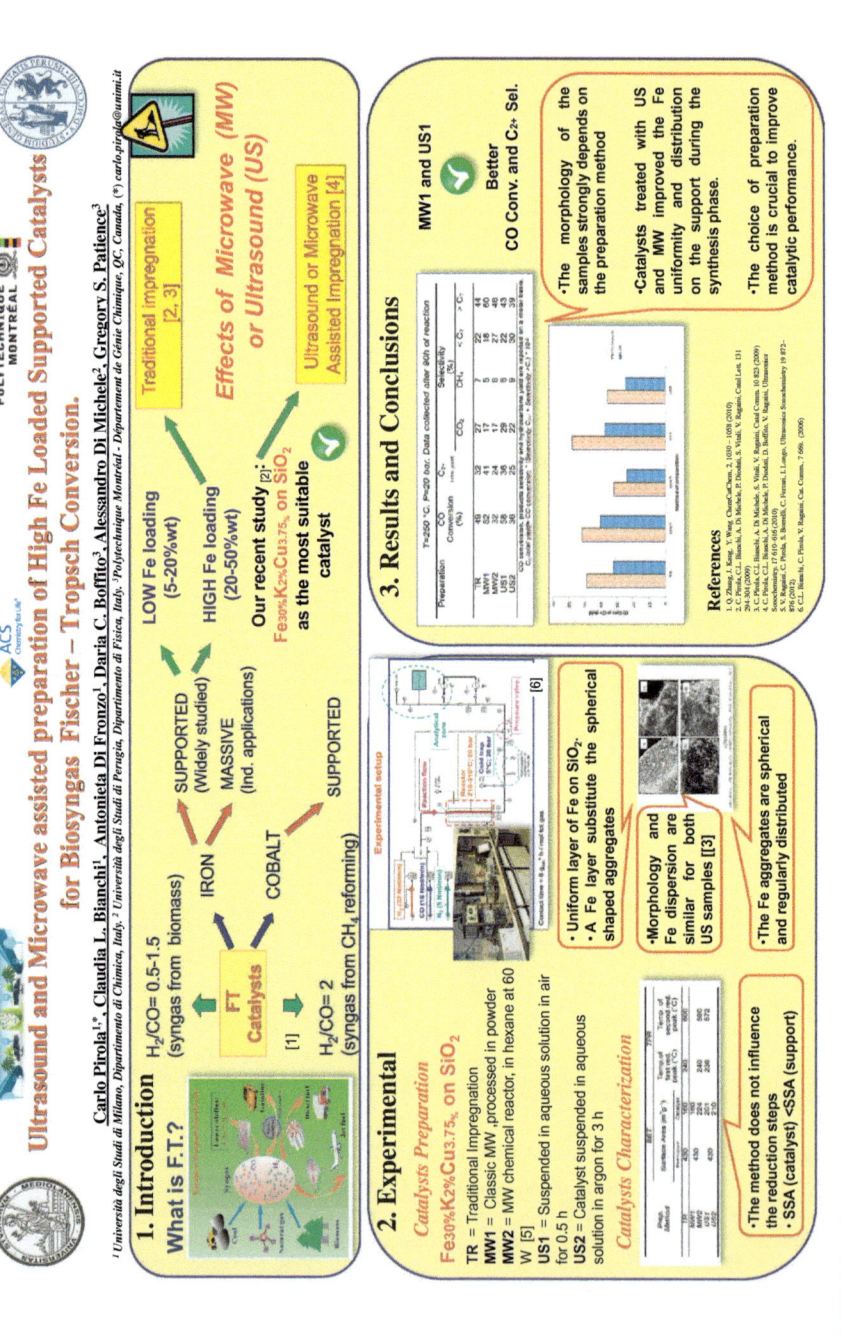

FIGURE 8.5   The "Introduction" panel spans the entire width and has a single image. The "Experimental" panel has three images and one table. The results and conclusions are grouped together (with references) and the panel has one table and one figure. The poster has 5 images and 2 tables, and fewer than 200 words. (From Pirola et al. (2013).)

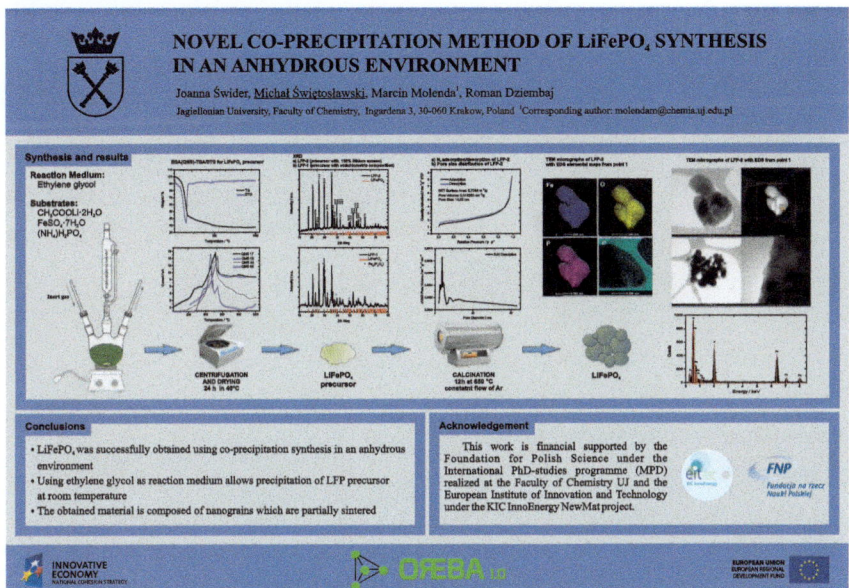

FIGURE 8.6   The composition of this three-panel poster is excellent. Swietoslawski et al. (2014) conduct the viewer from left to right, demonstrating each step of their work with images of the experimental equipment. The authors state conclusions and acknowledgments in panels below the experimental/results panel, which they label "Synthesis and results."

graphs that identify the physicochemical property that describes the principal results. They put a short title below each piece of equipment. With 50 words they demonstrate their work and give their conclusions. They don't explicitly state an objective, but this is not a significant flaw. The objective is implicit in the title, and since the conference was devoted entirely to battery materials, no one had a problem understanding the focus. At multidisciplinary conferences, state the context.

Separating posters into a few independent panels is good, but if you have more than four you risk crowding the poster. Commercial programs (e.g., Edraw, PowerPoint, Visio) offer standardized presentation templates that you can adapt to posters, such as flowcharts, business matrices, Venn diagrams, and fish bones. Adopting any of these formats helps focus the poster on images because there is little room to add blocks of text. You have to fit figures, tables, and graphs into geometric shapes. These shapes occupy most of the space on the page (Figures 8.7–8.9). You have to be creative to adapt them to scientific posters.

Poster Size

Respect the poster dimensions the conference organizers dictate on the event's website. Organizers set the size and orientation to fit on walls or poster stands. They account for the expected number of posters when choosing between

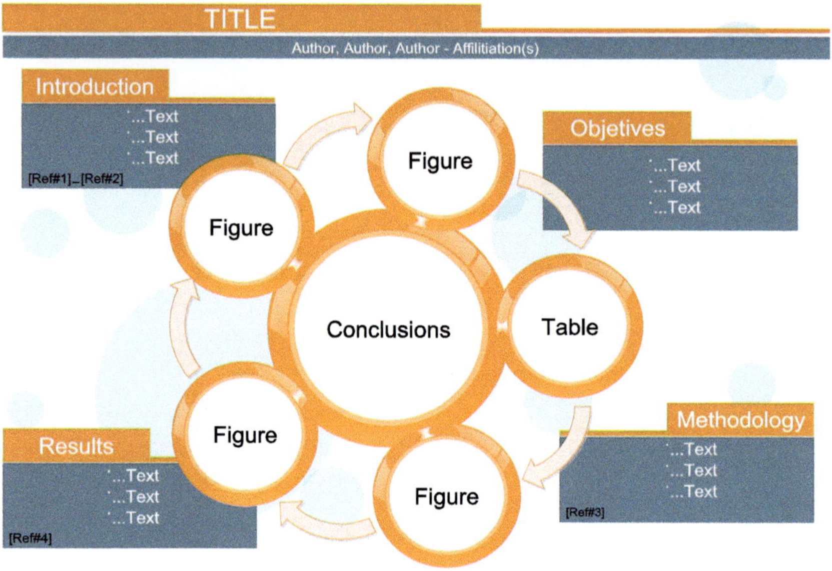

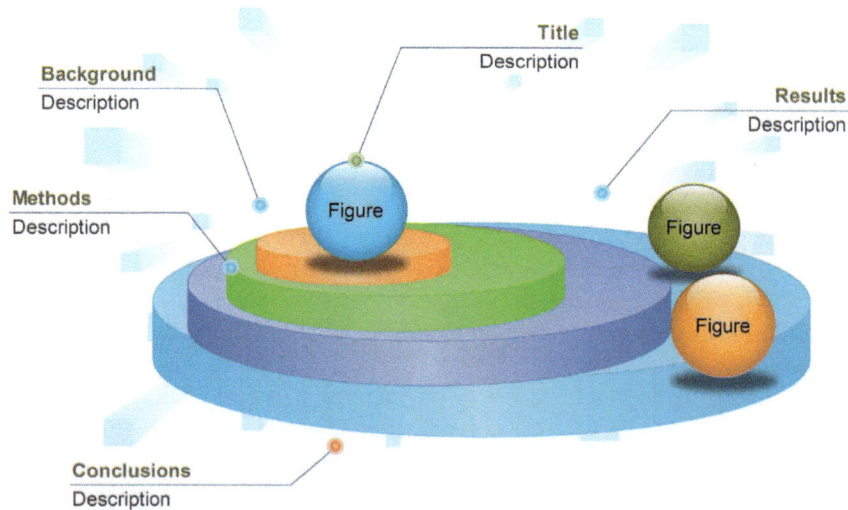

FIGURE 8.7   This flowchart poster combines text outside the circles, and images and tables inside them (modified Edraw image).

FIGURE 8.8   The white space and the simplicity of this poster format make it attractive (modified Edraw image).

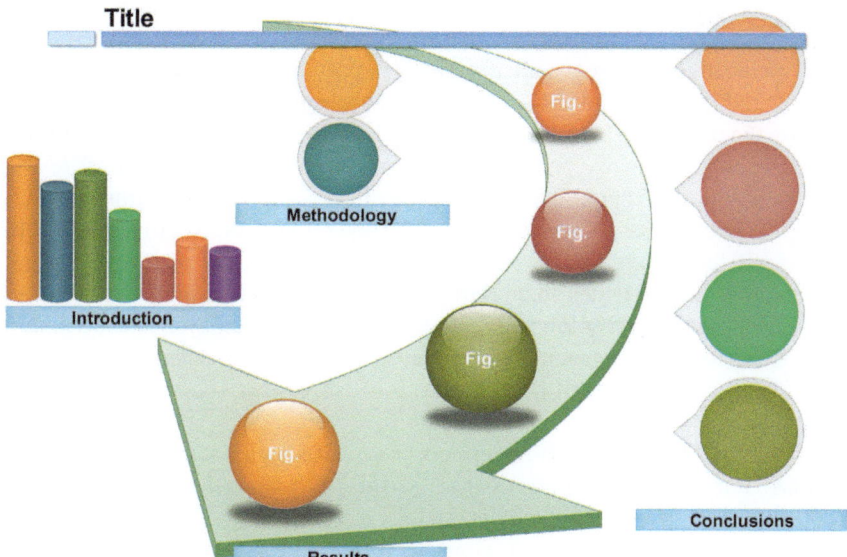

FIGURE 8.9   The arrow separates the sections elegantly (modified Edraw image).

vertical and landscape orientation, so follow their instructions for logistical reasons. You do not want your poster to protrude from the board.

If the organizers of the event make poster templates available, use them. This sounds obvious, but seldom are posters the same size at conferences even when a template is available. Sometimes organizers are at fault. They have an obligation to communicate what they want at least 1 month in advance. It is annoying to design a poster entirely and then to find out 1 week before the event that you have to change the format, so check with the organizers first. If you cannot get in touch with them, feel free to use your own layout. Adopt the standard poster size for North America or Europe (Table 8.4). Landscape orientation is always better than vertical orientation because most of the poster is close to eye level and it is easier to span data, figures, and text horizontally than vertically.

TABLE 8.4 Standard Poster Sizes

	Inches	Centimeters	Feet
North America	36 × 48	91 × 122	3.0 × 4.0
Europe	32 × 47	80 × 120	2.6 × 3.9
	39 × 55	100 × 140	3.3 × 4.6

Colors and Images

Contrast between the text and the background color is important. The gray background in Figure 8.6 is too dark. Avoid multiple colors for the background because the public can have trouble reading the poster. Stick to light backgrounds and dark fonts. (Light text on dark backgrounds can be attractive, but dark text on light backgrounds is better.) Gradient backgrounds—light at the bottom and dark on top, or vice versa—also work well. However, the darkest part of the poster should still be a pale color. Resist spending time deciding whether amaranth matches baby blue better than Barbie pink. Standard background styles are available together with color palettes (see Chapter 7). You may also use a texture or an image as a background—mountains, beaches, experimental equipment, etc. However, give yourself time to redo it because what you see on the screen does not always look good when printed.

Remember to select high-resolution images for your background and illustrations. For the background, do not go for anything below a resolution of 1000 px × 1000 px, to avoid the "pixelated" or blurry effect. Dots per inch (dpi) is the number of pixels per inch along an axis. It does not measure resolution. For instance, a 300 dpi image 3 in. long will have 100 px. When you resize it to 30 in, it will still have only 100 px, but they will be spread over 10 times the length. What matters for resolution is the total number of pixels with respect to the actual dimensions of the poster.

A dewdrop at a resolution of 2500 px × 2500 px is very clearly expanded to a 48 in. × 48 in. poster size, but when the picture is only 250 px × 250 px, it becomes blurry when expanded to the poster size (Figure 8.10).

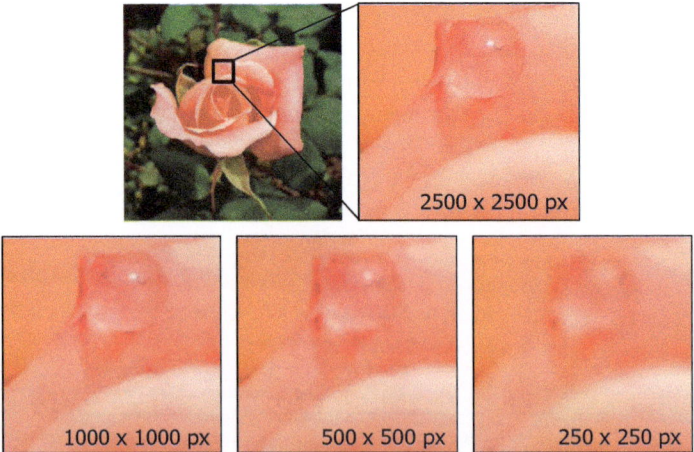

FIGURE 8.10   This rose was the background image of a 48 in. × 48 in. poster. The quality is excellent from images with a resolution of 1000 px × 1000 px and more.

Ensure that images you take from the Web are copyright free and royalty free and greater than 500 dpi. Stock.xchng, Iconfinder, OpenPhoto have royalty-free images for either commercial use or personal use.

Text

Size letters of the title so that people can read the poster from 3 m away. Title font sizes range from 72 pt to 110 pt. Authors' names are half that and the text in the poster ranges from 36 pt to 48 pt (Table 8.5). Put references at the bottom of posters, but limit them to 24 pt or less.

Limit the number of font sizes to four or less, and be consistent across the different categories: headers, labels, figures, tables, captions, text.

TABLE 8.5 The First Column Shows the Section Titles, With the Font Size in the Second Column

Title Section	Font Size, pt	Range, pt
Title	72	72-110
Authors	48	48-60
Affiliations	36	36-48
Subtitles	48	48-60
Text	36	36-48
Graph Text	28	24-32
References	20	16-24

Note: The third column gives font size ranges for each section.

Tables

Graphs are better than tables, and tables are better than text. Looking at tables takes more time than looking at graphs, and people are therefore less likely to do it. If you can't present the data with a graph (descriptive data or property data), limit the number of columns and rows, and highlight the most significant results—with color, bold, or font size (or all of these). Use arrows to relate cells in the table to graphs, conclusions, or discussions. We discourage shading table columns and rows in papers, but shading can be effective to highlight a row or column of text in posters and presentations. Follow the guidelines in Chapter 5 to format tables.

Graphs

Graphs are the focal point of posters. After reading the title, some people look at the introduction, but those familiar with your topic (those most likely to study your poster) will jump directly to the graphs to see how your results compare with what they know.

Well-crafted captions or titles ensure that your graphs are self-explanatory. Avoid repeating the axis titles in the caption. Pie charts, bar charts, and 3D graphs are less common in papers, but are effective in posters.

How tall should a graph or other image be compared with the height of the poster? Fixing the minimum dimension of images sets the maximum number of images the poster can have. The smaller the images are, the more you can fit. Set the shortest graph axis equal to 15% of the shortest poster dimension: for a 100 cm × 140 cm poster, the shortest axis of the graph is 15 cm. The length of the y-axis fixes all other dimensions: line weight, text font, and symbol size (Table 8.6).

TABLE 8.6 Elements of Graphs in Posters Are Proportionately Bolder Those of Than Graphs in Papers

Graph Element	Characteristic Dimension
y-axis length	150 mm
y-axis weight	3 pt
x-axis length	200 mm
x-axis weight	3 pt
Tick	3 mm × 1 mm
Axis title	32 pt
Tick label	28 pt
Legend label	28 pt
Symbol	24 pt

Note: The y-axis is the scaling dimension for characters and line weights.

8.3 PRESENTATION

Delivery

The poster should be on your left as you look at your audience. Stand straight and look confident. As you present the poster, move to the other side (since the flow of information is usually from left to right). Allow viewers to approach as close as they need to see the poster and hear what you are saying, so don't stand in the middle. Welcome people passing by but don't pester them. You might engage them by saying, "Would you like to hear about…" or "If you have any questions, I would be happy to answer them." Be natural and transmit the enthusiasm that you feel for your work.

Conference participants ask many types of questions: some of the answers are on the poster, some you know, some you never thought of. Your poster is a summary of your work and will not have every single data point that you measured. Bring supplementary materials, copies of your poster, and your contact information to distribute to people who are interested. Leave these handouts at the poster for people who don't get a chance to listen to you.

- Make eye contact periodically with each individual.
- Don't stare at your poster.
- Don't read your poster.

Questions and Answers

Answering questions in front of posters is less stressful than answering questions after an oral presentation. You have a better chance to explain yourself (especially after repeating yourself five or six times). You can listen to others' opinions and even admit you are wrong, which is easier in front of a poster than in a conference hall. Furthermore, you can keep asking the person to repeat the question until you understand exactly what that person wants to know.

Most people want you to review your work quickly so they can move on to the next poster, while others would like to hear the minutiae. In both cases, start by presenting yourself. Ask them for their name, give them your business card, and get theirs. This information will help gauge their interest and for how much time they want to listen. After the introductions, you have to convince them that you have the tools to address the problem. Then go backwards: state one or two of the most significant aspects of your work (conclusions), show one graph from the results that supports what you say, and finally show the equipment that you used to get to that point. Ask them if they would like more detail and if they don't want more, thank them for listening. For those who want to hear more, this is a good moment to start a conversation. Ask them what their interests are, and keep this information in mind as you proceed. Relax a little and take time to give them the general context: what makes your research interesting and what you want to accomplish. It is not a monologue so remember to ask questions too: "Did you notice this?" "Did you know that?"

Be polite, smile, have fun, and enjoy the questions. Some will be easy to answer, some you will have to think about, and others can be embarrassing:

- "Have you thought of doing it this way?"
- "What is your most significant result? Why?"
- "I don't understand your conclusions. Can you explain them to me in another way?"
- "Is that conclusion coherent?"
- "What is the uncertainty in your measurements?"
- "Are you familiar with the work of…?"
- "What are you planning to do next?"
- "I don't believe anything on your poster."
- "That violates thermodynamic principles."

The answers to some questions might address issues that go beyond the scope of your work. Others you might not have an answer to. In this case, do not make it up, but reply honestly.

- "I don't know."
- "You are right, we'll correct this in the manuscript."
- "Good point."
- "I am not sure that I understand your point."
- "We didn't do that, but it is an interesting thought."
- "Thank you for the question. It might be intriguing to investigate this subject."
- "I will look at that when I get back to the lab and I will let you know. Thank you for bringing it up."
- "We plan to investigate this subject in the next phase of our work."
- "We did not have the means to investigate that aspect."
- "Could you tell me specifically what you think is wrong with the data?"

Don't deviate from the question. Address it as well as you can. It's not an exam. If the question is precise and addresses a specific point, give a precise answer and don't ramble.

Poster sessions are the best opportunities to make contacts. People in your field are those most likely to come to your poster. Don't miss the opportunity to meet and impress them; they are likely to review your work for journals and are potential collaborators (and employers). Networking is very important to advance your career.

8.4 JUDGING CRITERIA

Poster competitions should focus not on the content, but on how it's communicated. If people walk away understanding what you did, you succeeded, whether or not you win the best poster award. Judges have to concentrate on elements for which they have objective criteria. Content is secondary. Our judging criteria (Table 8.7) focus on the poster and the presenter.

TABLE 8.7 Format and Presentation Weigh More Heavily Than Content Because Judges Can Compare These Elements Quantitatively

	Content
30	Originality Experimental rigor Compelling results
	Format
40	Layout Text Graphs Colors
	Presentation
30	Masters subject, follows poster Speaks clearly, enthusiastic Questions and answers

8.5 EXERCISES

1. *Poster design*: Identify positive and negative aspects of the posters in the following figures:
   a. Figure 8.11,
   b. Figure 8.12,
   c. Figure 8.13,
   d. Figure 8.14.
2. *Banners*:
   a. Improve the title of the poster in Figure 8.5.
   b. Improve the title of the poster in Figure 8.6.
   c. Identify positive and negative aspects for each banner in Figure 8.15.

Tungsten-Vanadium mixed oxides for the oxidehydration of glycerol into acrylic acid

F.Cavani[a], S. Guidetti[a], C. Trevisanut[a], G. Befi[a], L. Nieto[b], M. D. Soriano[b], P. Concepción[b]

[a]Dip. Chimica Industriale e dei Materiali, Università di Bologna, v.le Risorgimento 4, I-40136 Bologna and INSTM, Research Unit of Bologna, Italy.
[b]Instituto de Tecnología Química, UPV-CSIC, Campus de la Universidad Politécnica de Valencia, Avda. Los Naranjos s/n; 46022 Valencia (Spain)

Introduction

Glycerol is considered one of the Bio-based Building Blocks (B3) with the greatest potential for implementing and putting the biorefinery concept into practice. Among the several chemical compounds that can be produced starting from glycerol, one of the most studied routes is the transformation of glycerol into acrolein, while much less attention has been paid to its direct transformation into acrylic acid. The overall transformation formally includes a dehydrative step into acrolein first, requiring Brønsted-type acidity, and a second oxidation step into acrylic acid. The aim of the present work is to combine several chemical steps into a single catalyzed transformation using a single bifunctional catalyst.

Experimental conditions

Reactivity experiments were carried out in gas phase in a continuous flow reactor made of glass, operating at atmosphere pressure. The feed composition contained 2 mol% glycerol, 4 mol% oxygen, 40 mol% water, and 54 mol% helium. Temperature range was 290 – 410°C.

The Reaction

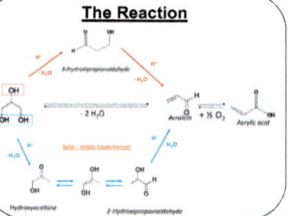

Catalysts preparation

V-free and V-containing hexagonal tungsten bronzes (HTB) were prepared hydrothermally. V-free tungsten-containing gel was prepared from an aqueous solution of $(NH_4)_6H_2W_{12}O_{40}\cdot H_2O$ with oxalic acid (W/oxalic acid molar ratio of 1/3), while V-W-containing gels were achieved from aqueous solutions of $(NH_4)_6H_2W_{12}O_{40}\cdot H_2O$ and vanadyl sulfate (and acidified with HCl until pH = 1.0) using several V/W ratios. The gels were loaded into Teflonned stainless steel autoclaves and heated at 175°C for 48h. The solid obtained was filtered off, washed, and dried at 100°C for 16 h. Lastly, the solids were heat-treated at 450°C (V-free sample) or 600°C (V-containing samples) for 2h in a N_2-stream. The catalysts have been called WV-n, in which n is the V/(V+W) ratio in the synthesis gel (in %).

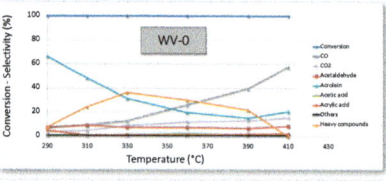

Characteristics of W-V mixed oxides catalysts

Sample	V/(W+V) ratio		S_{BET}	TPD[b]		Crystalline phase[d]
	In Gel	In catalysts[a]	(m2/g)	μmol NH_3 g-1	μmol NH_3 m-2	
WV-0	0	0	30,6	135	4,4	HTB
WV-2	0,17	0,12	19,0	72,3	3,8	HTB
WV-3	0,23	0,21	20,6	76,3	3,7	HTB
WV-4	0,29	0,23	23,8	107	4,5	HTB
WV-5	0,33	0,24	21,0	112	5,3	HTB
m-WO_3	0	0	2,0	23,7	11,9	Monoclinic

a) Catalyst compositions of calcined samples are obtained by EDX.
b) NH3-TPD results: amount of NH3 adsorbed in μmolNH3 g-1 or μmolNH3 m-2
c) H2-TPR results: H2-uptake (in mmolH2 g-1) in the 450-550°C (peak 1) and 550-750 (peak 2) temperature region
d) Hexagonal tungsten bronze (HTB).

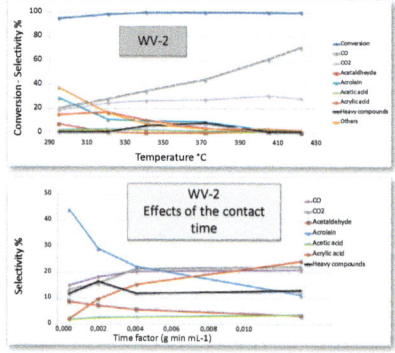

A comparison of the catalytic behavior of W-V mixed oxides catalysts.[a]

Sample	Sel_{max} Acrolein (%)	Sel_{max} Acrylic acid (%)	Sel CO_x (%) (T < 310°C)	Sel heavies (%) (T < 310°C)
W1	67	1	12	14
W1V0.2	29	18	39	< 5
W1V0.3	11	26	47	7
W1V0.4	8	18	54	5
W1V0.5	4	22	63	< 5

In all cases conversion is > 99%, but with WV-2 (conversion 98%). Catalyst amount 0.5 g (in granules), time factor time factor 0.0085 gcat min mL-1.

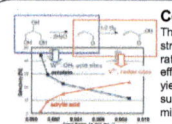

References:
F. Cavani, S. Guidetti, L. Marinelli, M. Piccinini, E. Ghedini and M. Signoretto, Appl. Catal. B 2010, 100, 197.

Conclusions:

The one-pot transformation of glycerol into acrylic acid can be carried out using W-V mixed oxide with the hexagonal bronze structure, with V4+ incorporated inside the WO3 lattice. The optimal ratio between the two elements corresponds to an atomic ratio V(W+V) in the heat treated samples of 0.12-0.21; this allows combining both the acid properties of W oxide, which is an efficient catalyst for glycerol dehydration into acrolein under aerobic conditions, and the oxidizing properties of V ions. The best yield to acrylic acid obtained was 25%. However, during lifetime experiments it was found that the progressive generation of surface V5+ species, which however occurred without any structural transformation of the hexagonal bronze into a monoclinic mixed oxide, led to a decrease in the selectivity to acrylic acid and to the concomitant rise in carbon oxide formation.

FIGURE 8.11   Tungsten-vanadium mixed oxides for the oxidehydration of glycerol into acrylic acid. (From Cavani et al. (2012).)

Activity Profile of Modified ZSM-5 for Methanol-to-Olefins

Mads Kaarsholm, Jesper Nerlov*, Roberta Cenni*, Jamal Chaouki and Gregory S. Patience

École Polytechniquede Montréal, C.P. 6079, succ. Centre-Ville, Montréal, Québec, H3C 3A7, Canada

* Haldor Topsoe A/S, Nymøllevej 55, 2800 Lyngby, Denmark

The product distribution and deactivation in the MTO process over a catalyst containing 10% H-ZSM-5 was studied in small diameter fixed bed reactors. Deactivation time of the catalyst is highly dependent on the residence time, doubling the feed rate decreased the deactivation time by a factor of ten and lowered the olefin production. Changing the feed form pure methanol to 10% methanol in nitrogen reduced methanol capacity of the catalyst significantly.

Introduction

Motivation for the MTO study

- Increased demand of olefins
- Propylene demand increases by 5%/year
- Prices doubled since 2003

General reaction pathway

$$2CH_3OH \xrightleftharpoons[+H_2O]{-H_2O} CH_3OCH_3 \xrightarrow{-H_2O}$$

$$C_2^= - C_5^= \longrightarrow \begin{array}{l} paraffins \\ aromatics \\ cycloparaffins \\ C_{6+} \ olefins \end{array}$$

In the present study, the effect of space velocity, temperature and reactor diameter on the activity profile of the catalyst is studied

Experiments

Catalyst

ZSM-5 (CBV 28014) Si/Al ratio of 140 imbedded in a Si/Al matrix by spray drying

Wetted with a $(NH_4)_2HPO_4$ solution and calcined to contain 1.5% phosphorous.

Catalyst size 400-600μm

Experiment

3, 6 and 8 mm inner diameter fixed bed reactors

- WHSV 0.22 – 2.38 h^{-1}
- Temperature 400 - 500°C
- Feed: 10% MeOH in N2, pure MeOH

Product analysis by GC (Hewlett Packard 5890 series-II)

Results

Visual deactivation of catalyst

Carbon deposition on the catalyst is clearly seen after a few minutes on stream

Catalyst deactivation can be divided in three zones

- Top: Light carbon deposition, initial reaction
- Middle: Heavy carbon deposition, main reaction
- bottom: Light carbon deposition, excess catalyst

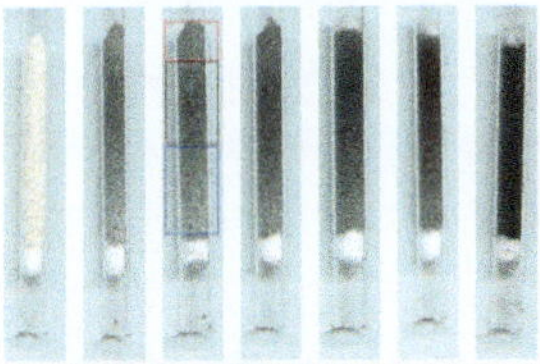

Catalyst coking during an experiment at time t = 0, 10, 20, 50, 110, 170, 1310 minutes in 5 mm reactor, 10% MeOH in Nitrogen, T=500°C, WHSV 0.43 h^{-1}

Results

Product distribution profiles of the MTO process

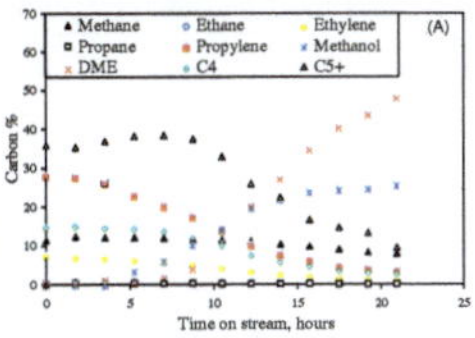

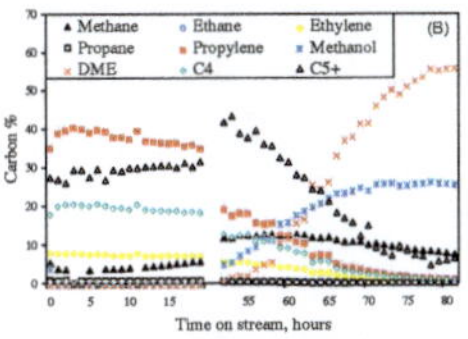

Product distribution at 500°C in 6 mm reactor with 10% methanol in nitrogen feed. (a) WHSV 0.43 h^{-1}, residence time 0.63 s (b) WHSV 0.22 h^{-1}, residence time 1.25 s

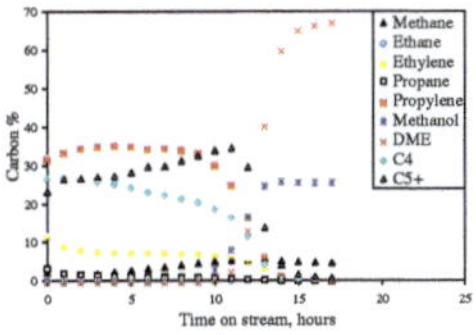

Product distribution at 500°C in 8 mm reactor with methanol feed. WHSV 2.38 h^{-1}, residence time 1.13s

- Product distribution highly dependent of WHSV
- Significantly increase in catalyst lifetime by reducing the WHSV from 0.43 h^{-1} to 0.22 h^{-1}
- Catalyst methanol capacity decreases with reduced partial pressure of Methanol
- Methane production increases at low partial pressure of methanol
- Reducing the partial pressure of methanol increases the selectivity to propylene on the expense of C$_4$ (at the same residence time)

Effect of temperature and reactor diameter

The increase in light olefin selectivity is clearly seen with increased temperature

Methane formation increases with temperature

Deactivation rate 400°C>>500°C>450°C

Slightly faster deactivation with reduced reactor diameter

A increase in light olefin selectivity was observed in the 3 mm reactor compared to the 6 mm

Reactor diameter, mm	3	3	3	6	6	6	6
Temperature	500	500	500	500	500	450	400
WHSV, h^{-1}	1.47	0.87	0.43	0.86	0.43	0.22	0.25
Residence time, s	0.18	0.31	0.63	0.32	0.63	1.25	1.24
Product distribution, %							
Methane	7.2	7.8	7.6	5.3	12	3.6	2.0	1.2
Ethane	0.3	0.3	0.4	0.2	0.8	0.2	0.1	0.1
Ethylene	0.9	3.1	7.3	2.0	7.5	8.3	5.4	2.7
Propane	0.1	0.1	0.6	0.1	0.4	0.9	0.6	0.1
Propylene	2.0	11	32	8.1	28	40	39	16
Methanol	16	22	1.0	14	0.6	0.1	0.0	17
DME	69	27	0.1	16	0	0	0.3	28
C$_4$	1.5	7.8	17	6.0	15	21	24	11
C$_5^+$	2.9	21	33	48	35	26	29	24

Product distribution after 2 hours on stream

Conclusion

- Carbon deposition on the catalyst is most severe a short distance down the catalyst bed
- Deactivation time is highly dependent on residence time doubling the feed rate (0.22 to 0.43 h^{-1}) decreases the deactivation time by 10
- Light olefin production increases with decreasing WHSV after complete oxygenate conversion is obtained
- The methanol capacity of the catalyst is highly dependent of the partial pressure of methanol, A significant decrease in methanol capacity is observed at low methanol partial pressure
- Decreasing the reactor diameter under similar process conditions increased the light olefin selectivity on the expense of a faster deactivation

FIGURE 8.12 Activity profile of modified ZSM-5 for conversion of methanol to olefins. (From Kaarsholm et al. (2006).)

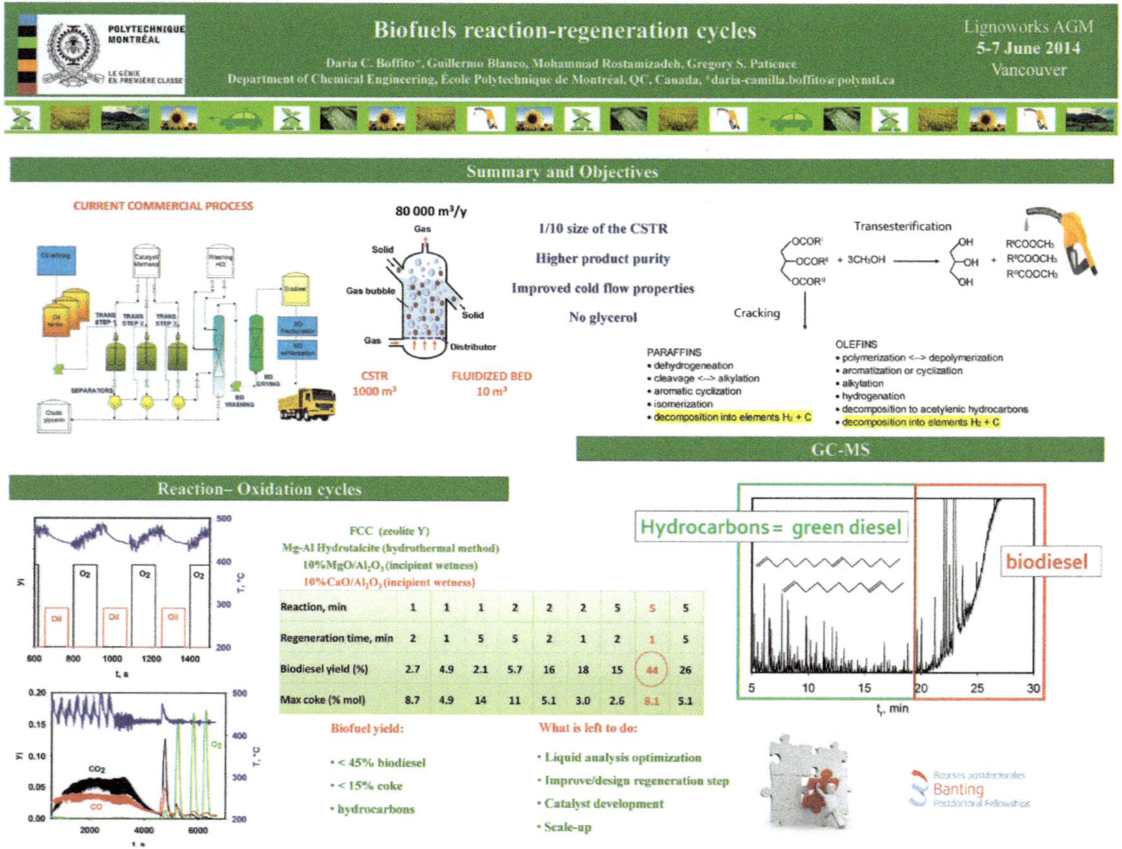

FIGURE 8.13 Biofuel reaction-regeneration cycles. (From Boffito et al. (2014).)

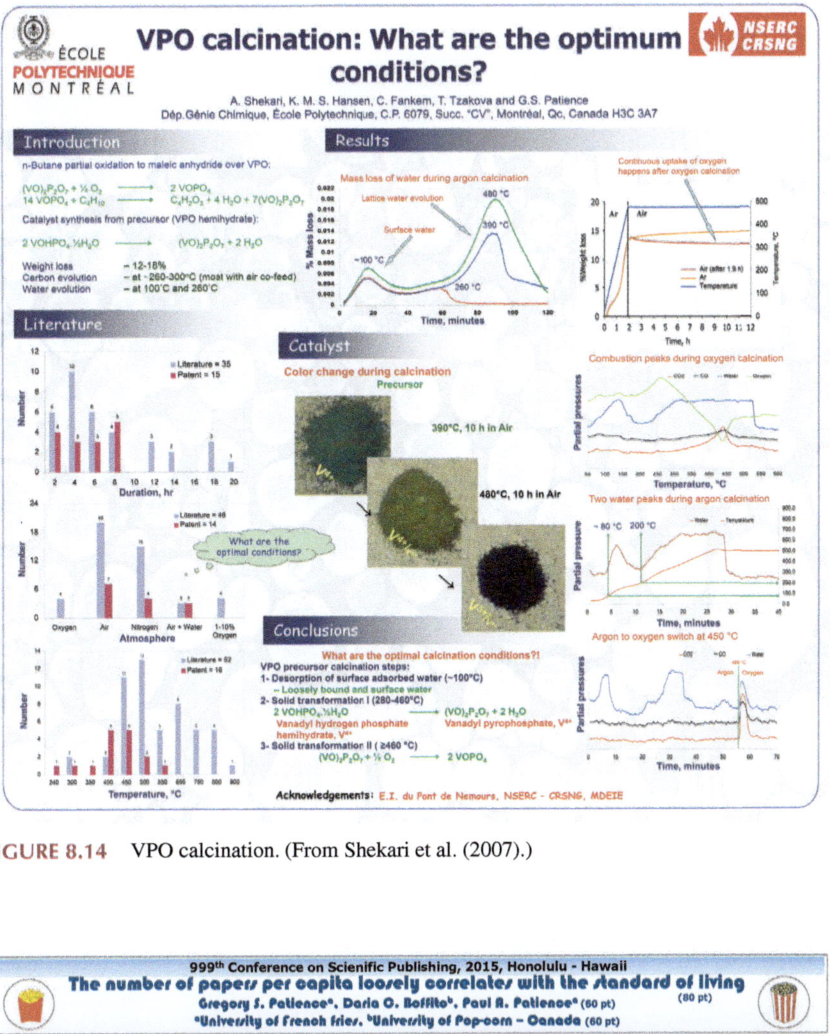

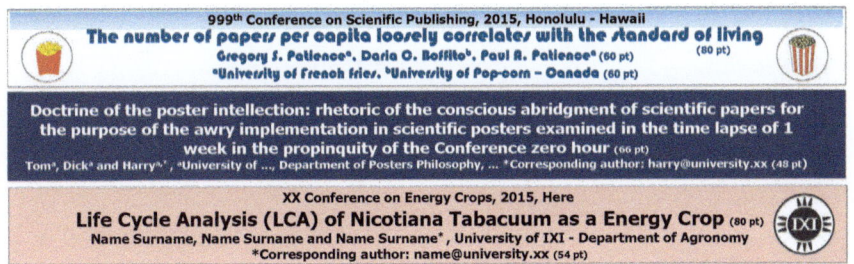

FIGURE 8.14   VPO calcination. (From Shekari et al. (2007).)

FIGURE 8.15   Examples of poster banners.

3. *Introduction*: List three suggestions to improve the blocks representing the introduction and experimental methods of the posters in the following figures:
   a. Figure 8.1,
   b. Figure 8.5,
   c. Figure 8.6.
   Improve the "Material and Methods" section at the following link: http://familymed.uthscsa.edu/rrnet/documents/StudentPosters/Lai%20 %20Poster.gif.
4. *Results and conclusions*: How would you improve the "Results and Conclusions" sections in the posters in the following figures:
   a. Figure 8.5,
   b. Figure 8.13.
5. *Graphs*: What is the ideal size for a graph on a 4 ft × 3 ft (width × height) poster?

Chapter 9

Plagiarism

Chapter Outline

Failing to properly attribute authorship of copied text, images, or ideas is plagiarism (Taylor, 2013). Obviously, copy-pasting substantial amounts of text is plagiarism, even if you cite it. Now, if you place quotation marks around the said copy-pasted text, it is a different story. Indent the text if there are more than four or five lines. (Refer to the journal's style guide.) Changing a few words from the copy-pasted text is still plagiarism. The motivation to plagiarize ranges from garnering false or undo credit (Naturephotonics, 2009) to sloth and the desire to increase one's perceived scientific output. Recycling work from an author's previous publications is perhaps a commoner source for plagiarists than copying the work of others.

The severity of the consequences of plagiarism increases with the level of responsibility of the plagiarist: teachers give high school students a zero in an assignment or exam; professors can fail undergraduate students taking their course or the institute can suspend them for 1 year or more (University of Toronto, 2014; Polytechnique, 2014); institutes may expel graduate students; employers can fire professionals; and universities can discipline professors and staff (restrict accessibility to research opportunities, for example). In January, 2013, the University of Düsseldorf withdrew the doctorate of Annette Schavan (the then German Minister of Education and Research) awarded in 1980 because she had plagiarized. In 2011, Karl-Theodor zu Guttenberg (the then German Minister of Defence) resigned after the University of Bayreuth revoked his doctorate. Baron Guttenberg admitted that he inadvertently copied text from other sources and put it in his thesis. The press was unkind to him and called him Baron zu Googleberg. Don't plagiarize: it is not worth the risk.

We include this chapter as part of scientific communication to demonstrate how plagiarism is related to intellectual fraud. Our first goal is to sensitize

Communicate Science Papers, Presentations, and Posters Effectively. http://dx.doi.org/10.1016/B978-0-12-801500-1.00009-7

science and engineering students and professionals to publishing ethics and establish a clear norm. The examples demonstrate what constitutes plagiarism and how to avoid it. We differentiate between plagiarism and copyright infringement with concrete examples. The exercises at the end reinforce the importance of publishing original work.

9.1 FORMS OF PLAGIARISM

Programs exist that detect plagiarism. Editors run them to check submitted articles against a database of papers. For example, 200 publishers add articles to the CrossCheck database, which now has more than 30 million articles (Elsevier, 2014a). CrossCheck reports a similarity index comparing an article with all the articles in the database. Deja vu is a database for MEDLINE citations (Errami et al., 2009). Errami et al. (2007) found that 1.35% of papers for which there was at least one common author (shared author) were very similar. For articles with no shared authors, 0.04% were very similar, which represents four papers out of 100 000.

Although it seems clear what constitutes plagiarism, there are nuances. Elsevier (2014a) classifies three levels of plagiarizing text:

1. literal copying
2. substantial copying
3. paraphrasing

The first one is easy to detect, but texts that are literally copied and pasted may be benign universal truths or phrases that thousands have quoted to contextualize the subject:

Carbon dioxide (CO_2) is the most abundant greenhouse gas .... This (incorrect) statement is repeated on the Internet over 40 000 times and 82 times in articles indexed by Google Scholar (2014). (In fact, water vapor is the most abundant greenhouse gas; Google Scholar (2014) has indexed 63 articles that state this fact.)

For the cases of substantial copying and paraphrasing, Elsevier (2014a) suggests that both the quality and the quantity of copying/paraphrasing are important to evaluate if the text has been plagiarized or not. It relates to the word order, to the structure, and to the sequence of ideas. Changing the word order of a single sentence from another article might be borderline acceptable, especially if it serves only to contextualize the subject. Changing word order in a series of sentences from the same paragraph of an article constitutes plagiarism.

Copying text is one form of plagiarism, but copying ideas, logic, and experimental techniques is another form that is more difficult to detect (Naturephotonics, 2009). Neglecting to cite previous work, whether or not it is intentional, can be consider intellectual plagiarism because the scientific community attributes the idea or work to the plagiarist rather than the person who first had the idea (Naturephotonics, 2009). The scientific community is negligent with respect to the patent literature because patents are seldom cited. A patent is granted to an original idea that has never been disclosed: inventors have to demonstrate that they had the idea first. Any article with a similar focus for which a patent exists necessarily comes after the patent. (Otherwise the patent office either would not grant the patent in the first place or would revoke the patent.)

9.2 COPYRIGHT INFRINGEMENT VERSUS PLAGIARISM

How does plagiarism differ from copyright infringement? Copyright means "the sole right to produce, reproduce, perform or publish a translation of the work" (Canada Minister of Justice, 2012). Taylor (2013) considers that copyright infringement relates to financial prejudice—failure to pay. Plagiarism, on the other hand, is an affront to the person(s) who created the original work. It violates the standard code of scholarly conduct like falsifying data, fabricating data (these could have criminal implications), neglecting to recognize deserving coauthors (even adding people who have not contributed), and sending a paper to multiple journals at the same time. Stealing ideas is also plagiarism. Although the origin of the word is from Greek and Latin meaning *kidnapper*, plagiarism is not a criminal offence. Copyright infringement is a criminal offence in most countries, and courts impose penalties on infringers: they have to pay for each item infringed, legal and court costs, and damages related to lost income.

To differentiate between plagiarism from copyright violations, Taylor (2013) presented four concrete examples:

1. Copying and pasting text and images from a published article into another article *violates the rights of the copyright holder* and *is plagiarism.*
2. Copying and citing sections of a paper or images from another paper is not plagiarizing but *infringes the copyright* if the copyright holder did not authorize the author(s) to reproduce the image(s).
3. Copying text from the public domain (text published 50-70 years after the last author died—depending on the country or patents) *plagiarizes the author(s)* but does not infringe copyright.
4. Publishing material with permission from the owner of the copyright and citing it is neither a copyright violation nor plagiarism.

Copyright excludes anyone other than the copyright owner from distributing the work, but there are a few exceptions: "fair use [dealing] for the purpose of research, private study, education, parody or satire" (Canada Minister of Justice, 2012). Fair use also pertains to criticism, but the work must be cited. Copyright works can be disseminated by individuals that don't hold the copyright

- for noncommercial purposes (as long as it does not prejudice the copyright holder), or
- if it is adequately cited, or
- if the individual infringing the copyright had reasonable grounds to believe that they were not infringing copyright.

Generic information in the introduction section of papers is protected by copyright, and copying it is plagiarizing. All figures and tables are property of the copyright holder. When you assign the copyright of your work to someone (e.g., a publisher), you no longer own it and you have no right to reproduce it without permission: it belongs to someone else. Original authors do maintain rights under fair-use rules and can reproduce the original figures in diagrams for presentations or educational purposes, for example. To reuse figures in another paper or book or even on websites requires permission from the copyright holder. However, experimental data from tables or experimental data derived from graphs is part of the public domain. You can use the data to produce your own tables and graphs, but if the graphs and tables are substantially the same as in the original work, that might constitute copyright infringement and/or plagiarism.

9.3 SELF-PLAGIARISM

Is copying paragraphs from one of your previous publications plagiarism? Yes. It is plagiarism and violates copyright. Cite previous work regardless of who published it. When you have published extensively in a field, reformulating the experimental methods section so that it is original is challenging. Avoid copying and pasting, and try to compose the text independently without looking at older articles. Some sentences will invariably resemble the previous work, but they should be sufficiently different to be original. If the experimental procedure/equipment is so common, consider referring to previous papers, and don't repeat all the information. There are only so many ways to state the operating conditions of analytical devices. Consider how Perret et al. (2014a) and Perret et al. (2014b) describe their X-ray diffraction procedure and measurements (we italicize original text)

Powder X-ray diffractograms were recorded on a Bruker/Siemens D500 incident X-ray diffractometer using CuKα radiation. Samples were scanned at 0.02° over the range $15° \leq 2\theta \leq 85°$. *Diffractograms were identified using the JCPDS-ICDD reference standards (CeO$_2$ (43-1002), monoclinic (37-1484) or tetragonal (50-1089) ZrO$_2$ and Au (04-0784)). The oxide* crystal size (d_{hkl}) was estimated using the Scherrer equation	Powder X-ray diffractograms were recorded on a Bruker/Siemens D500 incident X-ray diffractometer using CuKα radiation. Samples were scanned at 0.02° over the range $15° \leq 2\theta \leq 85°$. Diffractograms were identified using the JCPDS-ICDD reference standards for Co$_3$Mo$_3$N (89-7953) and Fe$_3$Mo$_3$N (89-7952). Crystal size (d_{hkl}) was estimated using the Scherrer equation.
Perret et al. (2014a, p. 129)	Perret et al. (2014b)

These texts are virtually identical except for the details of the solids. They have not prejudiced the owner of the copyright—the copyright holder is the same publisher, but is it plagiarism?

Let's apply the Elsevier (2014a) criteria of quality and quantity to establish whether the passages contain plagiarized text: the quantity is sufficient to be considered plagiarism (several sentences), but the quality is insufficient. These passages have no impact on the theme, conclusions, or originality of the work. Perhaps the authors could have referenced the previous article, but according to Elsevier (2014a) criteria, the text in the second column might not be plagiarism.

9.4 REFERENCES

To what extent are authors required to reference previous work? Do scientific principles established over 100 years ago need not be cited? It is unnecessary to cite definitions from a dictionary, common knowledge, or generally held facts. The top 50 papers published from 1989 until the end of 2014 (excluding seven review papers each with over 100 references) reference 37 papers on average. A sample of 500 scientific articles from 2008 referenced 15 500 articles, which is 31 references per paper.

A bibliography lists research that the authors consider pertinent to their work. Rarely is it exhaustive, and only in the case of review articles do they attempt to be exhaustive. It helps the reader to verify and substantiate the thesis of the work and look for other related articles. It includes only pertinent work and excludes trivial information that is held to be true by the specialists in the field. Considering that the top papers cite about 37 papers, that would be a good target for a paper. Soares and Thomas (2014) consider that an inadequate literature review could constitute plagiarism, so err on the side of citing more papers.

Wikipedia is a source of literature, but is rarely cited for several reasons: the authors are anonymous, the work is not peer reviewed, and anyone can change the information at any time, which makes it less reliable. Although the substance of Wikipedia is not cited, copying from Wikipedia constitutes plagiarism.

9.5 PATENTS

The US Patent and Trademark Office granted over 185 000 patents in 2008, which is about 10% of the documents archived by Web of Science^{TM} (2014). However, among the 3140 references in reviews, articles, letters, and reports in the December 2008 issues of *Science* and *Nature*, not one was a patent (Weaver and Barden, 2009). The top 500 articles of 2008 cite 48 706 documents, but only 138 of them are patents. (Many of these papers were review articles: 146 cited 100 papers or more. Excluding these papers, the rest averaged 37 references.) A random sample of 500 scientific articles from December 2008 cite 15 000 articles, of which 20 are patents. Patents are granted on the basis of novelty, the fact that what is described is not obvious, and that it has a commercial value. How is it that not one patent was cited by *Science* or *Nature* or so few patents are cited in the general scientific literature? People publish many papers in the areas in which patents were the first disclose the idea. Not citing the originating patents constitutes prima facie evidence of plagiarism, copying ideas without referencing. Piehler (2009) notes that patents are granted with respect to originality and not scientific validity. Furthermore, he holds that researchers may be discouraged from reading patents or citing papers that cite patents to minimize the risk of willfully violating intellectual property. Industrial researchers willfully examine the patent literature to keep abreast of the latest practical developments. They are required to understand it to be able to improve on the state of the art. Neglecting the patent literature is plagiarism (according to the criterion of Soares and Thomas (2014))! Patents are granted on the basis of new ideas and demonstration of those ideas. Copying ideas is plagiarism.

Since there is no copyright on patents, you should theoretically be able to copy the words directly into an article. You would have to cite it, because otherwise it would be plagiarism, but you wouldn't be infringing the copyright. This is really an academic argument, though, because putting quotes around it isn't difficult. Some patents copy other patents verbatim. They are referenced, but the words are not copyright.

9.6 AUTHORS' RESPONSIBILITIES

Shewan and Coats (2010) clearly identify authors' responsibilities when publishing. Elsevier (2014b) itemize the duties of authors, editors, reviewers, and publishers. Briefly, the corresponding author approves all coauthors and is responsible for the correctness of the manuscript. Approving the coauthors means that they have all made a substantial contribution to the research and to writing and proofing the document. The authors must confirm that the work is original and has not been submitted elsewhere. Furthermore, they certify prior work has been correctly cited.

Consequences

You jeopardize your professional reputation (and career) when you plagiarize. The consequences of plagiarism apply to everyone who had a part in the paper, whether they copied the text themselves or not. Consider papers that are withdrawn from a journal's website: The journal doesn't identify the individual who plagiarized. Everyone's name is displayed. (However, at times the editor or coauthors might identify the individual that copied the material (Taylor, 2013).)

Section 11 of the "Code of Conduct and Best Practice Guidelines for Journal Editors" COPE (2011) states that "Editors should not simply reject papers that raise concerns about possible misconduct. They are ethically obliged to pursue alleged cases."

What does "raise concerns" and "pursuing alleged cases" mean? In fact, the Committee on Publishing Ethics (COPE) suggests that institutions, not the editors, should be responsible for disciplining authors (Marcus and Oransky, 2011). COPE (2008) has developed flowcharts to investigate plagiarism. Editor first establish the extent of copying and then confront the authors with the evidence. They can obviously reject submitted papers or recommend major revisions. Editors can withdraw published papers that plagiarize. Further, they can replace the paper on the website with a statement saying who plagiarized from whom (Figure 9.1). In some cases, an editor can refuse future articles from that person for a certain period of time and can recommend that the publisher refuse articles for any journal. In the extreme, they could contact other publishers and recommend that they also prohibit articles from that person.

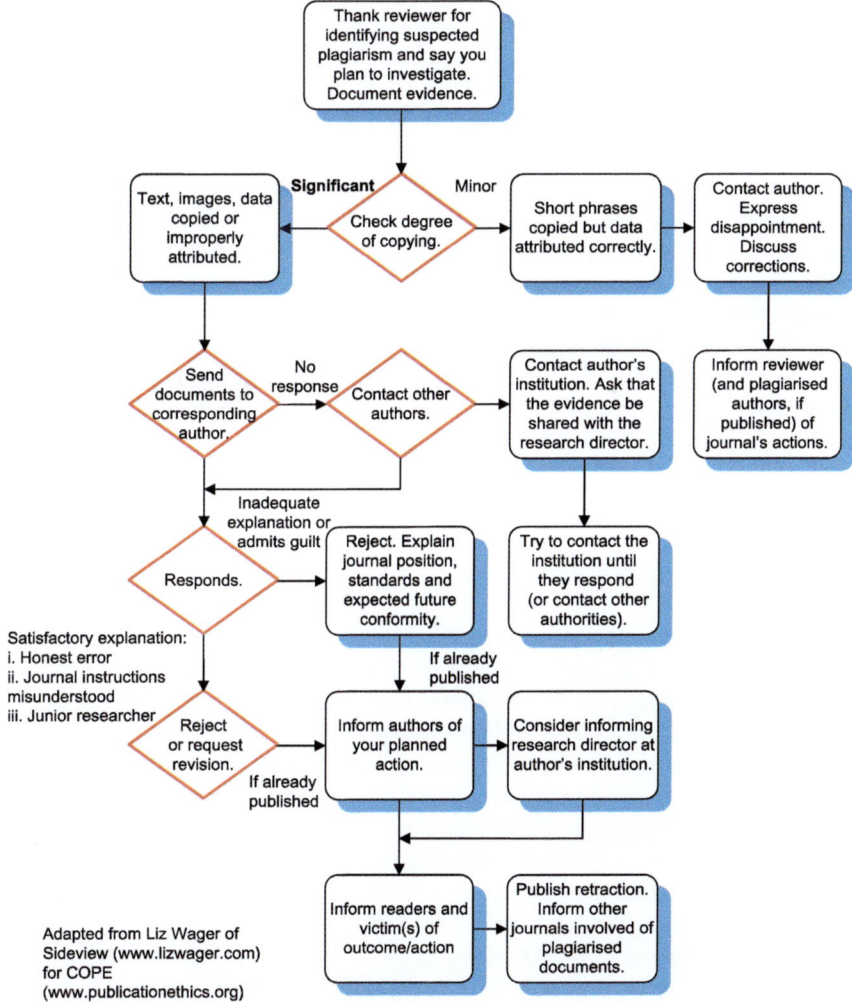

FIGURE 9.1 Flowchart for editors when a reviewer suspects that a document is plagiarized. (From COPE (2008), used with permission November 4, 2014.)

9.7 AVOIDING PLAGIARISM

The easiest way to avoid plagiarizing is to cite others! Don't copy-paste text then attempt to change a few words around in the hope that this will be enough to avoid plagiarism (cite the source). Begin by assembling related articles, read them, and jot down whatever you think may be important. Use bullet points. Bullets are very fast (ha ha). Make sure you write down the references next to where you wrote your notes, of course. Take a few days off, then compose text that you paraphrased from the original source. When composing text, put the articles away—avoid referring to them while writing.

Note that translating large sections of a document from one language to another (e.g., paragraphs) constitutes plagiarism. Translating sentences is fine as long as you cite the work.

If you are remotely unsure of whether what you want to do is permissible, ask for permission and cite it. In fact, it is important to verify every part of a collaboratively written document to make sure that no one has copied anything verbatim—on purpose or by accident. How are the collaborators supposed to know what has been copy-pasted and what is original? When a coauthor writes a couple of sentences very well but the rest of the text is riddled with grammatical errors, you have to wonder. Check it on the Internet—if the person was so lazy to copy and paste text, most likely that person was too lazy to get original work from the library. So, copy and paste the suspect text into an Internet search engine (and put the text inside quotation marks). This should either confirm your suspicion or allay your fear that the person copied the text verbatim. (Remember to check if the text is paraphrased.)

9.8 EXERCISES

1. Is changing a few words in a sentence enough to consider the new sentence original?
2. Is assembling a paper from a collection of other papers plagiarism?
3. To what extent can a student copy his or her own papers into a Ph.D. thesis?
4. Can you extract data from a graph and prepare a similar graph with this data and with other data without asking permission?
5. Does modifying an image (e.g., adding text) constitute original work? (Can you just put a picture between quotation marks?)
6. Can we copy-paste public domain images and text (e.g., text from patents)?
7. When does citing your own work become self-plagiarism?
8. Can you copy letters of recommendation that you write on behalf of someone (changing the name, of course)?
9. Can you copy text into grant applications or other documents that are not published? What if the text was from previous grant applications?
10. Can you copy a document format, such as that of a CV, for example?

11. How do you decide what is trivial or common knowledge and what information needs to be cited?
12. Many disciplines have a rich history of innovation. What proportion of recent papers versus authoritative older papers should be cited?
13. How many papers is considered sufficient to represent previous work in the field?
14. Can you copy the text of your patent directly into a scientific article?
15. Who has the copyright for sentences and paragraphs from company reports and patents?
16. Does assigning copyright to the publisher mean that you lose all rights to reproduce the ideas/text and images?
17. Weaver and Barden (2009) reported that the US Patent and Trademark Office issued 185 246 patents. Do you have to cite this reference?
18. The phrase "violation of the standard codes of scholarly conduct" appears 4210 times on the Internet. Is it plagiarism to reproduce this phrase exactly?
19. Many presentations posted on the Internet attribute the statement "if it wasn't published, it wasn't done" to E.H. Miller, 1993. Are the authors obliged to include the full citation? (We were unable to find the reference to Miller.) Most of these presentations cited both Miller and Whitesides (2004) on the same slide in the same order. Is this plagiarism?
20. Can an author translate and publish an article that has been published in another journal.

Appendix A

Solutions

Chapter Outline

A.1 CHAPTER 1: PUBLISHING INDUSTRY

1. Australia, New Zealand, Singapore.

3. As with question 2 there is no definitive answer. It takes as much as 100 h after all of the data has been collected and analyzed. The activities that take time include making graphs, preparing tables, formatting equations and text, organizing the structure, writing, proofing, and rewriting. Professional science writers allow 1 d to write 500 words. Papers might average about 3000 words, and most researchers are less proficient than *Science* writers. Assuming that we are half as proficient, writing a paper alone will take 6 d to 12 d or up to 100 h. It is easy to prepare graphs, but quality graphs can take hours (unless you have many excellent templates). Proofing and rewriting adds another day or two. So, even 100 h could be an underestimate.

5. Scientific productivity is a contentious issue, and any number we state is arbitrary and debatable. Some scientists have over 1000 citable contributions in the scientific literature. A few individuals' academic output is truly outstanding. We propose that producing six papers per year is good, and maintaining this rhythm over 40 years is excellent—240 peer-reviewed articles. The number of publications and the number of citations depend on the domain, but as a target, we suggest 4000 citations in a career would be excellent.

7. The median number of citations in Web of Science^{TM} (2014) is 10.

9. The first step to project the expected number of citations in 15 years is to calculate the asymptotic value, $\Phi(\xi)$, since we are given $\phi_{1,\xi}(5)$:

$$\phi_{1,\xi}(t) = \frac{11.3\xi^{0.45} \operatorname{csch}(0.053\xi)}{1 + \exp\left(\frac{6.2 - t\xi^{0.07}}{2.7}\right)}.$$

11. The impact of *CA: A Cancer Journal for Clinicians* is due entirely to the survey of cancer rates that are routinely cited several thousand times in the first couple of years. Excluding 27 articles of the 900 that this journal has published over the last couple of decades will reduce its IF substantially below 30. The IF should relate to the total number of citations. Journals that have thousands of articles that are cited infrequently have a greater impact than journals that publish fewer than 100 articles a year that are cited twice as frequently.

13. Many journals that publish review articles publish fewer articles per year, and so the h_5-index is moderate.

15. A journal's IF does not continually increase. It depends on the last 2 years. A researcher's h-index continually increases, but his or her h_5-index does not continually increase.

A.2 CHAPTER 2: WRITING STYLE

1. PdO/ZrO_2-Y is stable below 500 °C.

3. Directory-based cache coherency links connect the CPUs to minimize snoop transaction frequency.

    Swap the links and the CPUs and the sentence becomes active. The active sentence is two words shorter than the passive one because we remove the glue words *are* and *by*. They're present in the passive sentence only to make it grammatically correct. When you correct passive sentences, you often reduce the word count. Replacing the passive voice isn't the only change we can make.

5. We repeat the procedure until we reach sphere E.

7. The Apuseni Mountains are a remnant arc containing the more internal parts of the inner Dinarids.

9. Isoconversion methods are either type A or type B.

    The passive verb is *can be categorized*, and the patient is *isoconversion methods*. We can't identify an agent for this one, unless we can find whoever first thought of both groups. In fact, you're better off not even saying that so-and-so categorized isoconversion methods. You're better off stating facts: There are two categories.

11. Unix server Sparc Enterprise ran Sparc64 VI and Sparc64 VII.[7,8]

    Fujitsu called it Sparc Enterprise. Sometimes the subject doesn't follow *by*. The implicit agent in this sentence is the two references. They used Sparc64 VI and Sparc64 VII in Sparc Enterprise. But there is a better way to formulate this. Computers run operating systems. The true agent in this sentence is Sparc Enterprise, and the action is run. There is actually another passive verb in this sentence besides *were used*. It's *called*. This sentence is tricky because it contains two passive verbs. *Were used* is easy to catch. It's similar to the previous sentence, but instead of *by* we have *in*. The implicit agent is a person or group of people. We can't be sure who, and that's one problem with passive sentences. They sometimes hide the agent. If we add the agent, we neither add nor remove words. However, we add information.

13. These defence systems rely on small RNAs to detect and silence specific sequences of foreign nucleic acids.

15. Clays are also suitable acidic support materials, and we can add metals to tune their properties.

17. Oxford University radiocarbon-dated the collagen [3] from the adult mammoth by accelerator mass spectrometry. Birmingham University dated it by proportional counting of CO_2. (Coope and Lister, 1987, p. 474)

19. Icosahedral symmetry is compatible with quasiperiodic translational order, and states of matter arranged in such a way exist (i.e., quasicrystals[2]). (Cozzini and Ronchetti, 1996, p. 12040)

21. Helium is a critical tracer throughout the Earth sciences, where its simple isotopic systematics trace degassing from the mantle, date groundwater and time the rise of continents[1].

23. Here, we report that fusion fuel gains exceed unity.

Both *Science* and *Nature* encourage this construction in abstracts.

25. operating at 3 bar and 390 °C

27. The reaction rate was independent of flow rate from 20 nL/h and above.

29. Acid catalysts dehydrate xylose to furfural.

31. Quantify *extremely high*.

33. Heated in air

35. State what and how the phenomenon changes and quantify its value.

37. increasing temperature improves selectivity

The word *give* is a problematic verb and requires you to add other verbs to clarify what is going on. *Results* is nonspecific and you have to refer to a property. Here we have replaced *results* with *selectivity*. *Increasing selectivity* is desirable; *increasing results* has no meaning in this context.

39. represent

41. The new apparatus increased monoclonal antibody production rates by x%, which is satisfactory.

A.3   CHAPTER 3: REPORTING DATA

a. $d_p = 2.5\,\mu m$

Rather than quoting particle dimensions in meters, the text is reduced by using the correct prefix. The total number of characters is reduced from 11 to 6.

1c. $\$4000\,t^{-1}$ with a relative uncertainty $\Delta_\$ = 5\%$ or $4.0\,\$ \pm 0.2\,\$$

Economic analysis, cost estimates, and projections are inherently inaccurate. A detailed cost estimate is no better than 15% and cost overruns can exceed 100%. Unless the price of a product is provided directly by a vendor, stating a price with four significant figures is unwarranted.

1f. $\Delta H_f = -1500\,\text{kJ}\,\text{mol}^{-1}$

Physical parameters, such as heats of formation, derived from experimental data are inherently uncertain. Carrying seven significant figures is excessive. Reporting it as -1500 implies two significant figures and -1500.0 has four significant figures. For the case of $\pm 0.1\%$ uncertainty, report (Δ_x) explicitly.

1h. Ash content is $12\% \pm 1\%$. Measuring solids content is difficult: three significant figures might be too many.

1j. An increase of 39% and 36%

2. Identify acceptable expressions.

(d) and (e), but both are missing units. (d) states that the confidence interval lies between 7.98 and 7.88. (e) implies that the confidence interval is ± 0.005.

4. 3.060 from the correlation and 3.055 from the table.

A.4 CHAPTER 4: GRAPHS

1. Increase the x-axis font size to match the y-axis font size. You also have to apply bold typeface.

3. Identify each curve by placing an alphanumeric character beside it. If the curves intersect, you may need to assign different line types to each—we prefer solid lines for everything. The caption states that the line color corresponds to the symbol color in the legend of the insert. This is insufficient when the figure is copied in black and white.

5. (c) and (e). The line width in presentations should be heavier than in papers. Both (c) and (e) represent 2% of the plot size.

7. We discourage increasing either the x-axis range or the y-axis range to create space. Resist placing legends outside the plot area for papers. Graphic designers may reduce the plot size to accommodate the text. When there is too much text, put it in the caption. In presentations and posters, place the legends to the right of the plot.

9. (b) and (c). Text in figures should not more than 2 pt smaller than text in the main body. Sans serif is best, so choose Helvetica, Arial, Tahoma.... Both Helvetica and Arial font types are 1 pt larger than Times New Roman.

11. a. The caption text in *Nature Methods* most resembles 7 pt Trebuchet MS.

   b. This font size is equivalent to 7.5 pt Times New Roman. Text in graphs should be not more than 2 pt smaller than the caption but should be at least 6 pt.

    c. *Nature Methods* publishes many graphs with 5 pt Arial characters, which are uncomfortable to read. We recommend a minimum font size of 6 pt.

Science is equally inconsistent with font size even within the same article: the text in figures in Winzer et al. (2012) varies from 4 pt Arial to 8 pt Arial bold.

13. a. The legend is 3 pt, which is too small.
    b. The font size of the *y*-axis is much bigger than that of the *x*-axis.
    c. The plot is only 22 mm wide, which is very small.
    d. If this were not part of a multipanel graph, two data points would be insufficient to justify a graph.
    e. When the data differ by more than an order of magnitude, consider a logarithmic scale.
    f. They oriented the axis labels at $90°$, but for the other graphs in the paper the axis labels are oriented are at $0°$.
    g. The *y*-axis label states that the unit of measure is concentration, and then it is defined as % total alkaloids—this is unclear.
    h. The title *Capsules* is bold, while everything else is normal font.

15. The most powerful way to present multiple data sets on a single graph is to make multiple slides with the same frame: in the first slide of the series, put only one data set; when you advance to the second slide, add one set of data, and so on. Ya Huei (Cathy) Chin (University of Toronto) uses this concept to describe reaction mechanisms on a catalyst surface. She draws the catalyst surface and a compound, then describes the reaction, and demonstrates the interaction between a compound chemisorbed and a gas-phase compound, for example.

17. The quality of graphics from most analytical instrument computer packages is poor. They look good on the computer screen, but when they are printed, they are often illegible. You have to take the raw data and start from zero.

19. a. Increase the length of the *y*-axis. (The dimensions of the plot are small because the legend is large.)
    b. Place plot *(a)* on the left and plot *(b)* on the right.
    c. Delete the text beside *(a)* and *(b)* underneath the legend and place the letters in the top right hand corner of the plots. This will give extra space to increase the *y*-axis length. (The plot *(a)* caption begins one line below that of plot *(b)*.)
    d. Reformulate the legends and make a succinct table. Remove $d = x\,mm$, *Shell* temperature, *Tube temperature*, and *Dilution*.
    e. Place the legend in the plot.
    f. Rescale the plot *(b)* y-axis to 1200 K so that it is the same as the plot *(a)*.
    g. The label for plot *(a)* expresses the membrane thickness as δ, whereas in plot *(b)* thickness is d.

    h. Add a space before the % sign.
    i. Express the legend characters in normal typeface rather than bold.
    j. Replace the serif text with sans serif text.
    k. Choose other line types that are clearer.
    l. Color the lines to better distinguish them.
    m. Use the same sequence of line types.
    n. *Temperature* is unnecessary as an axis title.
    o. Express the y-axis in degrees Celsius (°C).
    p. Reformulate the caption. Replace *Effect of* .

A.5 CHAPTER 5: TABLE ESSENTIALS

1. The orders of magnitude of the variables are too different. Aligning the numbers at the decimal point would be ugly, so we leave them aligned to the right. The variables are like text, so we align them to the left. We set *interface* and *bulk* as subscripts to be similar to the subscript *Stefan* and to use less space. Options for the units include the SI convention (BIPM, 2006), where you divide the variable by the unit (Table A.1); and putting the units in a separate column (Table A.2). We put the units in a separate column to make the variable names clearer. You could remove the heading of the unit column, but empty column headings look incomplete, and could be confusing. Empty stub headings are fine.

    Though you can put the units column right after the stub, it reads more naturally when it's to the far right of the columns with the corresponding variables.

    We also reduced the number of significant figures from three to two.

3. Table A.3. The original table is full of units, which hides the values we're interested in. The first step is to organize the table into columns and to bring the units into the headers. The rate constants had three different units, so we made three columns. All the activation energies have the same unit,

TABLE A.1 Comparison between Velocity Perturbations and Stefan Velocity for Two Growth Conditions, $L = 10\,\text{cm}$

	Case 1	Case 2
$C_0'/(\text{m s}^{-1})$	130	130
P'/Torr	0.1	1
$D'/(\text{cm}^2\,\text{s}^{-1})$	120	15
$U_{\text{Stefan}}'/(\mu\text{m s}^{-1})$	4900	0.44
$U_{\text{interface}}'/(\mu\text{m s}^{-1})$	0.84	0.28
$U_{\text{bulk}}'/(\mu\text{m s}^{-1})$	7.5	7.5

TABLE A.2 Alternate: Comparison between Velocity Perturbations and Stefan Velocity for Two Growth Conditions, $L = 10\,cm$

	Case 1	Case 2	Unit
C_0'	130	130	$m\,s^{-1}$
p'	0.1	1	Torr
D'	120	15	$cm^2\,s^{-1}$
U_{Stefan}'	4900	0.44	$\mu m\,s^{-1}$
$U_{interface}'$	0.84	0.28	$\mu m\,s^{-1}$
U_{bulk}'	7.5	7.5	$\mu m\,s^{-1}$

TABLE A.3 Estimated Rate Constants. $k_t = k_{tc} = k_{td}$

	Pre-exponential factor		E_a
	s^{-1}	$m^3\,s^{-1}\,mol^{-1}$	$kJ\,mol^{-1}$
k_d	9.6×10^{15}		130
k_p		1.3×10^{11}	71
k_t		8.5×10^{18}	89
k_{tm}		2.9×10^{-2}	43
k_{ts}		4.5×10^3	68
$k_{tcd} = k_{tc}/k_{td}$		6.6×10^{-11}	61

so one column suffices. To scale the value of the activation energy, we changed the unit from "$kJ\,kmol^{-1}$" to "$kJ\,mol^{-1}$." To be consistent with this modification, we also changed the rate constant to "mol." Recognizing that the rate constants are derived from experimental data, we think it unlikely that the uncertainty is less than 0.1% (i.e., four significant figures). We kept two significant figures.

Table A.3 has a lot of white space because the pre-exponential factors have different units. To collapse the pre-exponential factors into a single column, we put their unit in a column (Table A.4). This contravenes our guideline that a column is a possibility you may consider for less than 2 s. In fact, it's a very bad idea. Be consistent. If units are going to be part of the rows, have a single column devoted to units. In this example, it's impossible. So, you should keep the units below the spanner heading. Do not do as in Table A.4.

5. The original table spans two columns in the journal. This is a waste of space for only five data points. Abbreviate the headings or add subscripts to variable names so the table is narrower. Since the table contains one catalyst

TABLE A.4 Alternate: Estimated Rate Constants (Revised).
$k_t = k_{tc} = k_{td}$

Rate constant	Pre-exponential factor	Unit	E_a kJ mol^{-1}
k_d	9.6×10^{15}	s^{-1}	130
k_p	1.3×10^{11}	m^3 s^{-1} mol^{-1}	71
k_t	8.5×10^{18}	m^3 s^{-1} mol^{-1}	89
k_{tm}	2.9×10^{-2}	m^3 s^{-1} mol^{-1}	43
k_{ts}	4.5×10^{3}	m^3 s^{-1} mol^{-1}	68
$k_{tcd} = k_{tc}/k_{td}$	6.6×10^{-11}		61

TABLE A.5 Amberlyst 46 Characteristics

SA m^2 g^{-1}	d_{avg} Å	V_{tot} mL g^{-1}	Acidity mequiv. H$^+$ g^{-1}	T_{max} °C
75	235	15	>0.4	120

and the caption states which catalyst it is, the stub is unnecessary and wastes space. Table A.5 would fit in one column of the journal. Parentheses around the unit are unnecessary when the units are all on the same line and the column heading text spans only one line.

A.6 CHAPTER 6: PAPER ESSENTIALS

1. *Title*:
   a. (i) Studies on, (ii) quantitative, (iii) qualitative, (iv) characterization;
   c. (i) Usage (repeated), (ii) model;
   e. (i) Analysis of, (ii) using, (iii) quantitative, (iv) method;
   g. (i) Definitions for, (ii) sepsis (repeated), (iii) guidelines, (iv) for the use of, (v) innovative;
   i. (i) Novel, (ii) methods, (iii) investigation;
2. *Abstract*
   a. We estimate the convection velocity of individual modes in turbulent shear flows on the basis spectral information in the temporal or spatial direction, while using local derivatives in other directions.
      Álamo and Jiménez (2009) want to say too much in the first sentence—49 words! We reduce it to 30 words, which is still too many.
   c. CuO splits water at 700 °C.

It is irrelevant to state in the abstract that they conducted experiments in a 7 mm ID quartz tube.

e. We have reproduced this abstract in it entirety. It is informative but we identify areas where the verbs could be more active.

> **First sentence** Lipid oxidation and microbial growth spoil nuts, fish, meats, whole milk powders, sauces and oils.
>
> > We combine the two causes of growth, eliminate *great variety*, and change *main cause of spoilage* to *spoil*
>
> **Third sentence** Adding antioxidants to the package and reducing the oxygen concentration are two strategies to limit lipid oxidation.
>
> > We change the word *some*, which is imprecise, to *two* because only two strategies were mentioned. Finally, we promoted the noun *addition* to a verb and eliminated one of the two *is limited*.
>
> **Fourth sentence** We propose active packaging that releases antioxidants continuously during storage.
>
> > We clarified who is proposing the idea with *we*, and eliminated *antioxidant* and *provide*.
>
> **Last sentences** We review methods to incorporate antioxidant compounds in the package, methods to evaluate their effectiveness and issues related to package design.
>
> > The two sentences (and our correction) are metadiscourse (saying what you are going to say)

f. The first sentence of the abstract of Hobaiter and Byrn (2011) is an example of self-conscious writing (Pinker, 2014). Self-conscious writers first have to justify their work, explaining how complicated and controversial it is.

- *Is known to be*—delete *known to be*.
- *There is controversy*—we should minimize *there is* and state what the findings are in the abstract.
- *Perhaps because*—again, leave speculation to the body of the report (it's best to substantiate speculation).

h.
- *We present here*—metadiscourse.
- *A method*—that becomes evident when you say what it is.
- *Analyzed separately*—again this is redundant when the authors show the analysis.

3. *Graphical abstract*:

a. The flowchart is illegible. The orientation is vertical rather than horizontal. The character size is too small. The quality is poor.

c. The thermotropic liquid crystalline polypeptide graphical abstract has a single image. Most of the space allotted to the graphical abstract is empty. The resolution of the image is poor, and the important elements are unclear. The authors have encircled four molecules, which is good, but they could have zoomed in to increase their resolution.

4. *Highlights*:
   a. Hu et al. (2015) repeat the title in the first two highlights. The third highlight is informative. The authors should add quantitative data in other highlights.
      - *We prepared*—this is metadiscourse and deviates our attention from the results.
      - *We examined*—a second example of metadiscourse.
5. *Introduction*:
   a. This sentence must be obvious to anyone who reads the *International Journal of Greenhouse Gas Control*. It has 33 words, which is too long. How is receiving *attention* going to change CO_2 emissions? The authors state that there is an effort to reduce greenhouse gas emissions and limit the increase—which one is it? Reducing the emissions is not the same as reducing the rate of increase of the emissions.
   c. As with the previous two examples, these two sentences are broad generalizations that contribute nothing. The third sentence of the introduction is specific and quantitative: *According to the International Diabetes Federation (IDF) statistics: the number of diabetes patients worldwide in 2011 has reached 366 million, an increase of nearly 30% compared with 285 million in 2010 (Ge and Xu, 2013).*
   e. $\theta = 3.1, m = 0.65$. Zhou et al. (2011) cite surprisingly few older articles. The modulus is very low, less than 1 (meaning recent articles are more prominent), and their scale parameter is only 3.1 years (versus 8 years for *Nature*). Both indices confirm that the authors virtually ignore older work. This could be because it is a completely new field or because they consider older work to be already in the historical record.
6. *Methods*:
   a. We prepared the Y_2O_3:Eu phosphor powder following the urea homogeneous method[2,7,18−21].
   c. The United States CDC enhanced its surveillance to identify people infected with influenza A viruses that could not be subtyped.
   e. We applied the Scherrer equation to calculate the crystallite size of pure LLDPE and DA-G/LLDPE nanocomposites that a D/Max 2500V/PC X-ray diffractometer measured (Rigaku Corporation, Tokyo, Japan) at room temperature.

      Or:

A D/MAX 2500 V/PC (Rigaku Corporation, Tokyo) measured the X-ray diffraction patterns. We applied the Scherrer equation to derive the crystallite size of pure LLDPE and DA-G/LLDPE nanocomposites.

$$D_p = \frac{0.89\lambda}{\beta \cos \theta}.$$

7. *Conclusions*:
   a. The first sentence is appropriate. In the second sentence, the authors talk about what they did rather than the results.
   c. **First sentence** It is obvious what this paper describes: the authors have stated it in the title, the abstract, and the body of the work.
   Second sentence The verb phrase *can be considered to provide a good representation of* is timid. The authors are frightened to take responsibility for their work. Instead, write "WFD represents/characterizes/predicts/...." The authors have mutilated the verb *represent*. What are *unreal* meteorological events?
   e. **First sentence** State right away the result.
   Second sentence The sentence is incorrect and could be shorter
   Third and fourth sentences The word *perform* appears four times in two lines. Consider shortening the sentence with "OK > RK > CK." The fourth sentence could be shorter.
   g. All the sentences are either trivial or lack substantive comments.
   First sentence Be quantitative.
   Second sentence This sentence could be omitted.
   Third sentence Fifty-eight words especially, with no punctuation, is too long. It is obvious that the findings come from the *present analysis*. A sentence one-third as long would convey the same message and be more readable too.
   Fourth sentence This sentence repeats the second one and could be shorter.

8. *Acknowledgments*:
   a. No. Analyzing data is necessary but insufficient to qualify as a coauthor. Coauthorship requires that you draft parts of the paper, re-read the paper, and accept responsibility for the paper.
   c. It is best to declare multiple affiliations to ensure potential conflicts of interest are evident. It is unnecessary to say that you are a company consultant if the company researchers are among the coauthors.

A.7   CHAPTER 7: PRESENTATIONS THEY WILL REMEMBER

1. *Introduction*:
   a. The colors are nice and there is white space, but the poster is incoherent.

   - Replace *Introduction* with *Light Olefin Market*.
   - Don't italicize.
   - Report the similar data in the same graph type (choose either a pie chart or a bar chart for both figures).
   - Abbreviate all legend labels, or none of them.
   - Remove the captions and title the graphs.

- Shorten the vertical axis title in Fig. 2 to *kt consumed*.
- Reduce the number of ticks to increments of 2000 kt.
- Remove *Ref#*. Put the reference and reduce the font size.

2. *Methods/experimental*:
   a. Too much detail.

   Good   The block diagram is an appropriate and clear choice to represent steps in chronological order.

   Good   Titles are good.

   Good   Page number and total number of pages are explicit.

   Change   There is too much detail. Substitute a schematic of your procedure for the experimental details. Take as an example Figure 8.6. Do not write anything that you are not going to mention.

   Change   Colors are faded.

3. *Results/reporting data*:
   a. The figures are clear and big in this slide.

   Change   Increase the font size of the title, which has the same font size as the text.

   Change   Reduce the font size of 50 μm, which is larger than the other text in the slide.

   c. This graph was copied from a software package.
   - Serif title—change to sans serif.
   - Label text too small.
   - Spelling mistakes in the *y*-axis title.
   - Tick labels inappropriate. (Figure 7.29).

4. *Conclusions*:
   a. Be quantitative—cite the data—and forget the qualifiers. (see Chapter 2). What does the speaker mean by *high* propylene selectivity? How *high* is the acidity? How long is a *long time*? The color scheme is poor.

5. *Colors*:
   a. The color palette in Figure 7.5 is monochromatic. A darker shade of green would be better with the white background.
   c. The labels are inconsistent. The first label has text and the second is a variable name. A split complementary color scheme or a monochromatic complementary color scheme would be better.

6. Find a subject that bores your audience (but not necessarily you).

7. A key element of this exercise is to assume your audience knows nothing and to concentrate on concepts from high school science. Keep the images uncluttered.

8. Remember to relax. It is almost finished. At best you can answer the questions brilliantly, at worst just mumble and ask for the next question. Whatever you do, be honest.
   a. Pause, take a breath. Count down from 3. Now that you might be a little calmer, reiterate what you think was a revelation. Perhaps you could be humble and say something nonthreatening such as *I am sorry you*

feel that way or *Perhaps, but we treated this subject in much greater detail.*

c. Respond with *I am very familiar with this excellent work. The difference between the previous studies and our work is....*

Or,

We are unfamiliar with this work and I would be happy to talk with you about it after this session finishes.

e. *We repeated all experiments three times and the standard deviation was less than 5%.*

g. *We calibrated our instruments once a week,*

i. *Excellent question, thank you so much. The process limitations relate to.... We believe that....*

k. *We evaluated the relative importance of all factors and these had least influence. Our next series of tests may well include these factors.*

A.8   CHAPTER 8: POSTERS THAT CAPTIVATE

1. *Poster design*:

a. Figure 8.11—standard format.

Good Title is clear and bold.

Good The number of graphs and tables is adequate.

Good The tables are well laid out and clear.

Change Write with bullet points to reduce the amount of text.

Change Brighten the slide with some color.

Change The legend text in the figures is too small.

c. Figure 8.13.

Good The balance between images and text is adequate.

Good The poster has sufficient white space.

Good The authors avoid mundane titles such as *Results and Discussion* and *Conclusions*.

Good Figures respect the criterion that the length of the y-axis be 20% of the poster length along the height.

Good The significant data in the table are in red and encircled.

Change Although the title is 76 pt, a larger font is better.

Change Remove the decorative images below the banner to give more space.

Change The poster needs more context even though the *Summary and Objectives* occupies 30% of the poster.

Change Replace *Summary and Objectives* as a label with something more pertinent.

Change Add experimental details.

Change Explain the graphs.

Change Remove the image of the man composing the puzzle to give more space.

2. *Banners*:
   a. Figure 8.5. The poster title is too long *Fe-supported catalysts for Fischer-Tropsch Process of Biosyngas* is better or *Microwave and Ultrasound Synthesis of Fe catalyst for Fischer-Tropsch*. A shorter title gives the banner more space, which may attract more visitors.
   c. Figure 8.15.
      i. Papers per capita
         Good The title is catchy.
         Good The contrast between the background and text is good.
         Good There is no address of the institution. Why do people include these addresses when anyone can search to find them on the Internet?
         Change Add the corresponding author contact information.
         Change Change the text type to sans serif.
         Change Increase the banner size so that the banner is less crowded.
      ii. Life cycle analysis
         Good The title is short (11 words) and informative.
         Good The title font size is good.
         Good The contrast between text and background is adequate.
         Good The banner contains all necessary information—logos, affiliations, etc.
3. *Introduction*:
   a. Figure 8.1.
      i. The pie charts report percentages for all of the compounds, but for highlighting only xylose is important. The pie sizes represent the percentages of the other compounds.
      ii. Too many different font sizes, uppercase letters, lowercase letters, bold versus plain text.
      iii. Too many bullet points—concentrate or three, maybe four (seems unbelievable when there are so many advantages).
   c. Figure 8.6.
      i. This poster is creative but the figures are too small. They are closer to 10% of the height than to 20%.
      ii. The authors could have used smaller frames around the sections and a narrower banner.
      iii. Text in the figures is too small.
      iv. A white background would increase the contrast.
4. *Results and conclusions*:
   a. Figure 8.5.
      - The first two columns in the table repeat the data in the graph. Omit them and either add series to the graph or include another graph with the data from the last three columns.
      - Increase the font size of the axis titles as well as the labels.
      - The bullet points are meaningless. Quantify the results or connect every bullet point to the numerical result in the graphs.

5. *Graphs*: The height of the graph should be 7 in. and the line weight of both axes should be 3 pt to 4 pt. The recommended tick size is 2.5 mm × 1.0 mm. The axis title font size should be 32 pt; tick labels, legend labels, and symbols font size should be 26 pt.

A.9   CHAPTER 9: PLAGIARISM

1. Yes and no. Strictly speaking, merely changing a few words is insufficient to constitute original work. However, if the *quality* or *quantity* of how much is copied is low, then perhaps it is not plagiarism. If a series of sentences that *represent* the idea of someone else are copied, then that is plagiarism (Elsevier, 2014a).

3. You can copy them entirely as long as you request and obtain permission from the copyright holder (publisher).

5. Modifying an image (e.g., adding text) might constitute original work. If the original image is still recognizable, ask the copyright holder for permission to reproduce it. You still have to cite it to avoid plagiarizing.

7. *Citing* your own work is never plagiarizing. *Copying* verbatim your own work might not be plagiarism either since you cannot kidnap yourself. If you copy something verbatim, you are probably infringing the copyright.

9. No. You should cite text from any published document but you can repeat sentences from your own grant applications (that have not been published) from previous years.

11. Newton and Galileo measured particle terminal velocities. This information is in the historical record. It is not trivial, but their measurements are inconsequential to the state of the art. Pertinent measurements, techniques, and ideas should be referenced. Avoid citing universal truths and dictionary definitions.

13. It depends on the subject but, on average, you should reference about 30 articles. The introduction could have between 10 and 20 references.

15. Patents are not covered by copyright. The assignee of a patent does not hold the copyright. Company reports are protected by copyright (owned by the company). However, reports of individuals are covered by copyright. It is plagiarism if someone copies the words from a report into an article without citing the author.

17. No. You do not have to cite Weaver and Barden (2009). The information they report is in the public domain. If you repeat a body of information that they report, then it is plagiarism if you don't cite them.

19. This is a gray area. Citing two people in the same order shouldn't constitute plagiarism or copyright infringement of the person who first cited those two people. However, if the format of the slide is almost identical—slide title, bold font, typographical errors—cite the original author. One Internet presentation cited the source, whereas others didn't. References should be accessible (Kamat and Schaltz, 2013).

Appendix B

Appendix

Chapter Outline

WEB OF SCIENCE CATEGORIES

TABLE B.1 Citations Ranked According to the Number of Papers Published Per Scientific Category (Web of Science) from 1989 to 2009 and 2009 to 2014 Together With $h_{5,2014}$ and $\phi_{25,0.1}$

Rank	Category	$N_{1989\text{-}2009}$	$N_{2009\text{-}2014}$	$h_{5,2014}$	$\phi_{25,0.1}$
1	Engineering, electrical, electronic	1 955 254	648 795	138	265
2	Biochemistry, molecular biology	1 567 630	372 960	268	815
3	Materials science, multidisciplinary	1 297 622	519 316	276	379
4	Chemistry, multidisciplinary	1 098 676	352 829	324	611
5	Physics applied	1 095 648	369 827	243	394
6	Neurosciences	959 935	276 359	196	589
7	Pharmacology, pharmacy	884 108	260 680	161	371
8	Oncology	844 221	312 322	229	522
9	Surgery	836 048	273 839	113	275
10	Chemistry, physical	828 835	273 716	266	527
11	Cell biology	818 179	207 689	264	926
12	Optics	776 335	229 230	125	249
13	Medicine, general, internal	756 732	211 133	308	1016
14	Clinical neurology	741 150	257 355	156	378
15	Physics, condensed matter	697 416	166 278	238	479
16	Cardiac, cardiovascular systems	659 942	217 197	175	470

Continued

Communicate Science Papers, Presentations, and Posters Effectively. http://dx.doi.org/10.1016/B978-0-12-801500-1.09979-4

TABLE B.1 Citations Ranked According to the Number of Papers Published Per Scientific Category (Web of Science) from 1989 to 2009 and 2009 to 2014 Together With $h_{5,2014}$ and $\phi_{25,0.1}$—Cont'd

Rank	Category	$N_{1989-2009}$	$N_{2009-2014}$	$h_{5,2014}$	$\phi_{25,0.1}$
17	Immunology	651 294	186 841	203	613
18	Environmental sciences	618 360	217 462	149	346
19	Computer science, artificial intelligence	607 950	171 982	103	319
20	Computer science, theory and methods	555 395	197 139	69	211
21	Telecommunications	539 158	159 998	89	224
22	Physics, multidisciplinary	529 138	146 566	190	662
23	Multidisciplinary sciences	528 885	201 926	343	1866
24	Radiology, nuclear medicine, medical imaging	521 233	155 367	115	342
25	History	511 944	113 488	18	30
26	Public environmental occupational health	511 074	180 358	113	366
27	Endocrinology metabolism	506 244	188 535	77	475
28	Engineering, chemical	501 415	159 492	128	255
29	Engineering, mechanical	496 474	144 272	146	165
30	Computer science, information systems	488 954	156 821	88	238
31	Gastroenterology, hepatology	487 186	147 277	140	345
32	Biotechnology, applied microbiology	480 960	170 055	187	535
33	Genetics, heredity	480 747	130 051	216	765
34	Plant sciences	479 038	136 005	123	504
35	Psychiatry	476 969	147 380	132	444
36	Hematology	469 345	137 911	138	416
37	Mathematics, applied	460 387	156 763	85	225
38	Astronomy, astrophysics	454 408	120 834	170	578
39	Peripheral vascular disease	446 292	120 841	148	528
40	Medicine, research, experimental	443 250	141 985	160	658
41	Chemistry, organic	442 677	117 042	117	298
42	Mathematics	422 156	135 821	65	151
43	Chemistry, analytical	412 447	121 387	119	336
44	Instruments, instrumentation	404 983	90 860	83	200
45	Microbiology	402 376	117 784	154	502
46	Economics	397 563	135 566	87	409
47	Computer science, interdisciplinary applications	395 456	130 349	120	359
48	Pediatrics	386 549	121 387	89	245
49	Biophysics	385 768	102 604	126	496
50	Geosciences, multidisciplinary	378 084	116 935	106	300

TABLE B.1 Citations Ranked According to the Number of Papers Published Per Scientific Category (Web of Science) from 1989 to 2009 and 2009 to 2014 Together With $h_{5,2014}$ and $\phi_{25,0.1}$—Cont'd

Rank	Category	$N_{1989-2009}$	$N_{2009-2014}$	$h_{5,2014}$	$\phi_{25,0.1}$
51	Mechanics	367 510	116 291	80	267
52	Physics, atomic, molecular, chemical	363 960	92 550	111	464
53	Biology	362 106	94 455	107	388
54	Urology, nephrology	356 094	118 307	106	285
55	Energy, fuels	355 712	153 821	143	238
56	Polymer science	352 522	98 179	128	348
57	Automation, control systems	350 747	114 566	94	265
58	Veterinary sciences	346 524	99 483	62	139
59	Food science, technology	345 061	118 796	83	265
60	Metallurgy, metallurgical engineering	342 657	99 694	73	226
61	Physiology	327 998	73 806	115	471
62	Engineering, civil	327 520	114 503	83	157
63	Computer science, software engineering	322 259	81 001	64	217
64	Ecology	318 904	96 386	138	569
65	Obstetrics, gynecology	312 657	104 573	85	230
66	Pathology	309 956	91 033	101	319
67	Nanoscience, nanotechnology	302 395	170 642	242	514
68	Chemistry, inorganic, nuclear	280 731	74 686	101	299
69	Engineering, biomedical	276 267	96 698	107	330
70	Zoology	271 962	79 456	67	262
71	Physics, particles, fields	269 604	68 859	120	458
72	Infectious diseases	268 786	89 470	122	344
73	Biochemical research methods	263 554	96 361	158	652
74	Computer science, hardware architecture	259 037	76 091	61	259
75	Respiratory system	257 038	94 327	106	370
76	Ophthalmology	256 572	57 127	81	270
77	Humanities, multidisciplinary	250 554	55 966	16	22
78	Nuclear science technology	250 112	56 616	53	167
79	Toxicology	248 170	75 960	81	371
80	Engineering, multidisciplinary	244 364	78 887	60	192
81	Education, educational research	235 457	93 516	60	172
82	Chemistry, applied	229 924	74 592	87	296
83	Dermatology	229 231	82 111	70	196
84	Management	228 360	86 255	82	601
85	Agronomy	224 853	52 171	63	212

Continued

TABLE B.1 Citations Ranked According to the Number of Papers Published Per Scientific Category (Web of Science) from 1989 to 2009 and 2009 to 2014 Together With $h_{5,2014}$ and $\phi_{25,0.1}$—Cont'd

Rank	Category	$N_{1989-2009}$	$N_{2009-2014}$	$h_{5,2014}$	$\phi_{25,0.1}$
86	Information science, library science	224 472	48 833	54	184
87	Water resources	224 105	69 650	77	256
88	Political science	220 238	54 832	47	177
89	Psychology, multidisciplinary	217 943	63 191	79	641
90	Engineering, environmental	217 187	79 905	113	332
91	Meteorology, atmospheric sciences	212 116	63 803	103	485
92	Dentistry, oral surgery, medicine	212 021	53 323	66	216
93	Physics, mathematical	211 543	63 393	99	385
94	Chemistry, medicinal	209 065	83 480	92	302
95	Marine, freshwater biology	208 405	60 325	80	287
96	Nutrition dietetics	206 414	70 426	100	380
97	Imaging science, photographic technology	202 408	40 706	70	291
98	Geochemistry, geophysics	202 217	60 371	86	440
99	Spectroscopy	201 938	51 168	73	306
100	Operations research, management science	201 777	71 846	66	274
101	Transplantation	197 131	75 808	103	215
102	Engineering, manufacturing	188 079	76 543	51	126
103	Psychology	187 685	53 910	102	660
104	Electrochemistry	184 210	75 198	114	357
105	Orthopedics	183 085	63 867	79	325
106	Crystallography	183 041	50 967	79	305
107	Physics nuclear	181 504	40 567	84	284
108	Religion	181 346	45 223	18	37
109	Health care sciences, services	176 566	73 388	84	328
110	Agriculture, dairy animal science	175 211	43 144	55	192
111	Business	173 377	56 393	76	633
112	Physics, fluids, plasmas	172 628	52 987	75	351
113	Statistics, probability	169 818	52 375	105	695
114	Sport sciences	169 087	61 804	81	304
115	Critical care medicine	168 137	63 460	106	396
116	Literature	164 215	31 825	16	27
117	Rheumatology	163 603	54 320	91	237
118	Materials science, coatings films	162 056	40 796	67	279
119	Engineering, industrial	159 940	48 697	52	151
120	Sociology	157 573	40 505	55	298
121	Engineering, aerospace	155 844	24 526	31	120
122	Virology	155 657	46 918	110	399

TABLE B.1 Citations Ranked According to the Number of Papers Published Per Scientific Category (Web of Science) from 1989 to 2009 and 2009 to 2014 Together With $h_{5,2014}$ and $\phi_{25,0.1}$—Cont'd

Rank	Category	$N_{1989-2009}$	$N_{2009-2014}$	$h_{5,2014}$	$\phi_{25,0.1}$
123	Reproductive biology	154 906	49 108	77	256
124	Psychology, clinical	151 835	46 894	90	413
125	Psychology, experimental	151 362	47 177	94	507
126	Philosophy	149 691	42 448	24	69
127	Acoustics	149 060	34 699	75	222
128	Thermodynamics	148 539	50 145	79	222
129	Anesthesiology	146 561	34 163	80	309
130	Language, linguistics	141 715	35 636	27	120
131	Oceanography	136 784	35 698	73	356
132	Materials science, ceramics	135 215	32 631	46	234
133	Mathematics, interdisciplinary applications	133 911	50 222	65	443
134	Behavioral sciences	133 729	38 450	99	429
135	Entomology	133 162	33 200	59	223
136	Construction, building technology	131 003	55 322	51	124
137	Remote sensing	130 622	36 879	63	302
138	Rehabilitation	129 617	48 397	62	205
139	Music	124 035	24 149	18	37
140	Anthropology	121 824	34 570	47	168
141	Law	120 274	29 628	37	193
142	Developmental biology	118 649	32 414	121	821
143	Literary reviews	118 624	21 015	5	10
144	Art	116 900	23 058	13	23
145	Otorhinolaryngology	116 596	33 446	49	166
146	Health policy services	116 399	47 359	69	308
147	Nursing	115 268	52 935	40	133
148	Materials science, characterization, testing	112 970	21 849	31	88
149	Agriculture, multidisciplinary	112 669	41 572	55	278
150	Environmental studies	112 116	42 081	72	214
151	Geriatrics, gerontology	111 559	37 451	79	280
152	Evolutionary biology	111 432	38 414	111	810
153	Social sciences, interdisciplinary	109 921	39 843	46	182
154	Horticulture	109 511	33 898	43	201
155	International relations	106 313	26 686	40	201
156	Medical laboratory technology	103 955	25 613	68	280
157	Fisheries	102 909	28 579	51	233
158	Area studies	100 219	24 658	20	54
159	Allergy	99 453	34 032	55	278
160	Literature, romance	98 161	21 119	5	9
161	Forestry	97 972	26 284	52	217

Continued

TABLE B.1 Citations Ranked According to the Number of Papers Published Per Scientific Category (Web of Science) from 1989 to 2009 and 2009 to 2014 Together With $h_{5,2014}$ and $\phi_{25,0.1}$—Cont'd

Rank	Category	$N_{1989-2009}$	$N_{2009-2014}$	$h_{5,2014}$	$\phi_{25,0.1}$
162	Soil science	97 935	23 310	61	308
163	Gerontology	97 412	37 893	59	266
164	Robotics	96 356	31 673	43	192
165	Business finance	94 903	31 233	58	391
166	Linguistics	91 162	34 340	44	227
167	Mathematical, computational biology	90 033	40 117	119	796
168	Psychology, developmental	88 827	28 524	76	466
169	Education, scientific disciplines	88 591	31 815	56	149
170	Geography	86 548	25 000	53	225
171	Planning development	86 077	21 329	48	227
172	Materials science, composites	85 704	24 235	56	189
173	Computer science, cybernetics	84 853	16 473	49	204
174	Parasitology	81 032	29 106	98	215
175	History philosophy of science	80 459	24 038	25	129
176	Psychology, applied	78 204	23 433	64	464
177	Transportation science technology	77 778	31 927	45	154
178	Substance abuse	77 155	29 027	62	287
179	Psychology, social	71 344	20 942	64	656
180	Geography, physical	70 405	26 773	70	312
181	Materials science, biomaterials	69 856	33 775	103	474
182	Mining, mineral processing	68 867	19 538	39	143
183	Social issues	68 025	14 744	34	147
184	Architecture	67 327	17 796	32	12
185	Geology	66 751	15 228	50	286
186	Engineering, petroleum	64 526	14 229	24	58
187	Communication	64 501	20 539	45	204
188	Tropical medicine	63 955	23 347	59	171
189	Emergency medicine	63 626	25 318	50	179
190	Social sciences, biomedical	63 543	20 391	52	264
191	Engineering, geological	62 457	19 224	32	148
192	Biodiversity conservation	60 219	23 996	80	394
193	Asian studies	60 056	12 829	9	23
194	Archaeology	58 593	14 251	11	121
195	Medical informatics	57 733	18 513	53	365
196	Materials science, paper, wood	57 592	11 056	36	105
197	Classics	57 367	11 600	7	28
198	Psychology, biological	55 904	14 618	54	387
199	Paleontology	52 331	16 610	41	238

TABLE B.1 Citations Ranked According to the Number of Papers Published Per Scientific Category (Web of Science) from 1989 to 2009 and 2009 to 2014 Together With $h_{5,2014}$ and $\phi_{25,0.1}$—Cont'd

Rank	Category	$N_{1989-2009}$	$N_{2009-2014}$	$h_{5,2014}$	$\phi_{25,0.1}$
200	Public administration	52 073	15 915	32	135
201	Mineralogy	51 429	15 171	53	390
202	Medieval renaissance studies	51 009	13 303	6	18
203	Film, radio, television	49 652	12 485	16	48
204	Ethics	47 309	16 102	33	126
205	Social work	46 604	13 338	34	187
206	Agricultural engineering	46 303	21 071	78	300
207	Transportation	45 986	15 990	39	153
208	Social sciences, mathematical methods	45 357	15 668	51	751
209	Urban studies	45 282	13 103	38	195
210	Family studies	44 730	13 475	42	272
211	Anatomy, morphology	44 109	11 337	45	255
212	Psychology, educational	42 960	13 767	52	467
213	Neuroimaging	41 908	13 890	91	673
214	Audiology, speech language pathology	41 766	12 004	44	296
215	Limnology	41 595	10 980	48	365
216	Mycology	41 322	12 803	49	308
217	Microscopy	41 224	9010	40	238
218	Women s studies	40 974	12 632	29	163
219	History of social sciences	40 585	10 181	16	55
220	Ergonomics	38 692	8235	32	182
221	Materials science, textiles	37 493	14 571	41	140
222	Engineering marine	36 363	6367	20	19
223	Criminology, penology	34 695	11 773	36	193
224	Hospitality, leisure, sport, tourism	34 221	18 322	37	184
225	Medicine, legal	33 817	10 694	43	165
226	Literature, German, Dutch, Scandinavian	33 751	5725	4	11
227	Theater	33 586	6837	8	18
228	Engineering, ocean	33 254	9863	32	64
229	Industrial relations, labor	32 680	6809	27	182
230	Education, special	32 050	9253	43	172
231	Integrative, complementary medicine	31 843	16 725	41	188
232	Primary health care	31 219	11 457	31	136
233	Agricultural economics policy	28 482	5551	27	125
234	Ornithology	28 331	6564	28	197
235	Cell tissue engineering	27 769	17 516	117	413
236	Demography	23 944	6300	31	261
237	Poetry	23 818	3735	4	11

Continued

TABLE B.1 Citations Ranked According to the Number of Papers Published Per Scientific Category (Web of Science) from 1989 to 2009 and 2009 to 2014 Together With $h_{5,2014}$ and $\phi_{25,0.1}$—Cont'd

Rank	Category	$N_{1989\text{-}2009}$	$N_{2009\text{-}2014}$	$h_{5,2014}$	$\phi_{25,0.1}$
238	Dance	23 325	3630	5	7
239	Psychology, psychoanalysis	22 393	4823	15	109
240	Literary theory, criticism	19 835	4731	5	25
241	Literature, British Isles	19 360	3628	6	21
242	Folklore	18 293	3780	5	23
243	Literature, American	17 780	3479	5	19
244	Ethnic studies	16 929	5497	23	98
245	Medical ethics	16 524	6643	27	106
246	Psychology, mathematical	16 062	3342	41	558
247	Cultural studies	14 690	8079	24	177
248	Logic	14 395	6160	16	136
249	Literature, African, Australian, Canadian	14 089	2676	5	16
250	Andrology	13 545	4122	33	200
251	Literature, Slavic	12 373	2977	3	7

WEB OF SCIENCE ARTICLES PUBLISHED BY COUNTRY

TABLE B.2 Country Population (in 2013), Number of Papers Published from 2010 to 2014 $N_{2010\text{-}2014}$, and Rank According to the Number of Papers Published Per Year from 2010 to 2014 ($\dot{N}_{2010\text{-}2014}$) Per 1000 Persons

Rank	Country	$N_{2010\text{-}2014}$	$\dot{N}_{2010\text{-}2014}$ per 1000	Population
1	Vatican	94	23.50	800
2	Switzerland	152 203	3.83	7 952 000
3	Scotland	87 064	3.29	5 300 000
4	Iceland	5202	3.25	320 000
5	Denmark	84 494	3.03	5 580 000
6	Sweden	131 964	2.78	9 490 000
7	Australia	293 948	2.57	22 902 000
8	Netherlands	215 247	2.56	16 838 000
9	Liechtenstein	459	2.55	36 000
10	Monaco	456	2.53	36 000
11	Norway	63 390	2.53	5 009 000
12	Singapore	62 359	2.41	5 184 000
13	Finland	63 296	2.34	5 408 000
14	Ireland	50 909	2.22	4 588 000
15	New Zealand	48 083	2.17	4 434 000

TABLE B.2 Country Population (in 2013), Number of Papers Published from 2010 to 2014 $N_{2010\text{-}2014}$, and Rank According to the Number of Papers Published Per Year from 2010 to 2014 ($\dot{N}_{2010\text{-}2014}$) Per 1000 Persons—Cont'd

Rank	Country	$N_{2010\text{-}2014}$	$\dot{N}_{2010\text{-}2014}$ per 1000	Population
16	Canada	375 517	2.15	35 000 000
17	England	594 843	2.12	56 000 000
18	Belgium	117 015	2.12	11 036 000
19	Slovenia	20 605	2.00	2 057 000
20	Austria	82 987	1.96	8 451 860
21	Wales	30 297	1.95	3 100 000
22	Israel	74 278	1.89	7 870 000
23	Caribbean Netherlands	191	1.81	21 133
24	Luxembourg	4500	1.76	512 000
25	Northern Ireland	14 677	1.63	1 800 000
26	United States	2 500 752	1.58	317 000 000
27	Germany	600 094	1.49	80 500 000
28	Greenland	413	1.45	57 000
29	Portugal	73 543	1.40	10 487 000
30	Estonia	9177	1.39	1 318 000
31	Spain	313 462	1.36	46 185 000
32	Taiwan	157 723	1.35	23 293 593
33	Cyprus	5601	1.34	839 000
34	Czech Republic	68 904	1.31	10 512 000
35	Greece	67 380	1.25	10 787 000
36	France	407 822	1.25	65 350 000
37	Italy	361 538	1.22	59 464 000
38	South Korea	269 341	1.11	48 580 000
39	Croatia	20 493	0.96	4 291 000
40	Grenada	477	0.86	110 821
41	Lithuania	12 811	0.80	3 190 000
42	Serbia	27 926	0.78	7 120 000
43	Japan	478 492	0.75	127 300 000
44	Slovakia	20 277	0.74	5 445 000
45	Hungary	36 411	0.73	9 962 000
46	Poland	126 691	0.66	38 501 000
47	Malta	1245	0.60	418 000
48	Romania	53 567	0.56	19 043 000
49	New Caledonia	709	0.55	256 000
50	Bermuda (UK)	168	0.53	64 000
51	Latvia	5422	0.52	2 070 000
52	Qatar	4048	0.45	1 792 000
53	Palau	41	0.41	20 000
54	Malaysia	57 830	0.39	30 017 000
55	Turkey	143 338	0.38	74 724 000
56	Chile	32 685	0.38	17 402 000
57	Bulgaria	13 183	0.36	7 284 000
58	Seychelles	160	0.35	90 945
59	St Kitts and Nevis	89	0.34	52 000
60	Guadeloupe	688	0.34	402 000
61	Tunisia	18 035	0.34	10 673 000

Continued

TABLE B.2 Country Population (in 2013), Number of Papers Published from 2010 to 2014 $N_{2010-2014}$, and Rank According to the Number of Papers Published Per Year from 2010 to 2014 ($\dot{N}_{2010-2014}$) Per 1000 Persons—Cont'd

Rank	Country	$N_{2010-2014}$	$\dot{N}_{2010-2014}$ per 1000	Population 2013
62	Iran	127 565	0.33	76 800 000
63	Montenegro	1007	0.32	620 000
64	French Guiana	362	0.32	224 000
65	French Polynesia	435	0.32	270 000
66	Barbados	415	0.30	273 000
67	Saudi Arabia	38 083	0.28	27 137 000
68	Lebanon	5722	0.27	4 228 000
69	Uruguay	4198	0.26	3 251 000
70	San Marino	40	0.25	32 000
71	Armenia	3937	0.24	3 275 000
72	Macedonia	2465	0.24	2 057 000
73	Dominica	82	0.23	71 000
74	Kuwait	3776	0.23	3 328 000
75	Argentina	45 450	0.23	40 117 000
76	Brunei	470	0.22	423 000
77	Réunion (France)	906	0.22	816 000
78	South Africa	55 725	0.22	50 586 000
79	Brazil	207 630	0.22	192 400 000
80	Russia	153 026	0.21	143 056 000
81	Oman	2870	0.21	2 773 000
82	Jordan	6225	0.20	6 297 000
83	United Arab Emirates	8095	0.20	8 264 070
84	China	1 206 309	0.18	13 473 500 010
85	Trinidad and Tobago	1073	0.16	1 318 000
86	Bosnia and Herzegovina	2853	0.15	3 840 000
87	Martinique (France)	287	0.15	390 000
88	Bahrain	898	0.15	1 234 000
89	Fiji	595	0.14	861 000
90	Georgia	3045	0.14	4 469 000
91	Belarus	6116	0.13	9 460 000
92	Ukraine	27 428	0.12	45 426 000
93	Thailand	37 957	0.12	65 500 000
94	Costa Rica	2454	0.11	4 302 000
95	Botswana	1152	0.11	2 038 228
96	Mexico	63 078	0.11	112 336 000
97	Panama	1806	0.11	3 406 000
98	Federated States of Micronesia	54	0.10	103 000
99	Andorra	39	0.10	78 000
100	Tuvalu	5	0.10	10 000
101	Egypt	39 173	0.096	82 019 000
102	Jamaica	1156	0.085	2 706 000
103	Moldova	1494	0.084	3 560 000

TABLE B.2 Country Population (in 2013), Number of Papers Published from 2010 to 2014 $N_{2010\text{-}2014}$, and Rank According to the Number of Papers Published Per Year from 2010 to 2014 ($\dot{N}_{2010\text{-}2014}$) Per 1000 Persons—Cont'd

Rank	Country	$N_{2010\text{-}2014}$	$\dot{N}_{2010\text{-}2014}$ per 1000	Population 2013
104	Albania	1147	0.081	2 832 000
105	Cuba	4533	0.081	11 241 000
106	Mauritius	510	0.079	1 286 000
107	Colombia	18 438	0.078	47 015 000
108	Gabon	583	0.077	1 505 000
109	Vanuatu	97	0.075	257 000
110	Bahamas	128	0.072	354 000
111	Mongolia	990	0.071	2 780 000
112	Gambia	590	0.068	1 728 000
113	Algeria	12 005	0.065	37 100 000
114	Marshall Islands	17	0.063	54 000
115	Belize	92	0.059	313 000
116	Morocco	9391	0.058	32 548 000
117	Azerbaijan	2562	0.056	9 111 100
118	Namibia	573	0.050	2 283 000
119	Antigua and Barbuda	22	0.050	88 000
120	Tonga	24	0.047	103 000
121	India	279 198	0.046	12 105 700 010
122	Bhutan	164	0.045	721 000
123	Venezuela	6046	0.045	27 150 000
124	Kenya	7305	0.038	38 610 000
125	Swaziland	223	0.038	1 186 000
126	St Vincent and Grenada	18	0.036	100 892
127	Guyana	135	0.034	787 000
128	Cameroon	3312	0.034	19 406 000
129	Pakistan	31 306	0.034	184 000 000
130	Kazakhstan	2773	0.033	16 734 000
131	Cape Verde	79	0.032	492 000
132	Libya	1006	0.031	6 400 000
133	Suriname	82	0.031	525 000
134	Peru	4599	0.031	30 135 000
135	Sri Lanka	3151	0.031	20 653 000
136	Senegal	1948	0.030	12 855 000
137	Solomon Islands	82	0.030	554 000
138	Ecuador	2120	0.029	14 483 000
139	Benin	1286	0.028	9 100 000
140	Republic of the Congo	561	0.028	4 043 000
141	Ghana	3332	0.028	24 223 000
142	Malawi	1781	0.027	13 102 000
143	Uganda	4332	0.026	32 939 000
144	Zimbabwe	1555	0.024	12 754 000
145	Maldives	40	0.024	330 000
146	Saint Lucia	20	0.024	167 000
147	Vietnam	9670	0.022	87 840 000

Continued

TABLE B.2 Country Population (in 2013), Number of Papers Published from 2010 to 2014 $N_{2010\text{-}2014}$, and Rank According to the Number of Papers Published Per Year from 2010 to 2014 ($\dot{N}_{2010\text{-}2014}$) Per 1000 Persons—Cont'd

Rank	Country	$N_{2010\text{-}2014}$	$\dot{N}_{2010\text{-}2014}$ per 1000	Population 2013
148	Lao People's Democratic Republic	679	0.022	6 256 000
149	Bolivia	1131	0.021	10 624 000
150	Zambia	1373	0.021	13 046 000
151	Guinea-Bissau	155	0.020	1 521 000
152	Kiribati	10	0.020	100 000
153	Burkina Faso	1511	0.019	15 731 000
154	Iraq	3068	0.018	33 330 000
155	Nepal	2317	0.017	26 621 000
156	Tanzania	3722	0.017	43 188 000
157	Papua New Guinea	578	0.016	7 014 000
158	Kyrgyzstan	447	0.016	5 477 000
159	Nigeria	12 593	0.016	158 423 000
160	Syria	1681	0.016	21 566 000
161	São Tomé and Príncipe	12	0.015	165 000
162	Cambodia	971	0.014	13 396 000
163	Paraguay	436	0.014	6 382 000
164	Uzbekistan	1984	0.014	29 123 000
165	Philippines	6148	0.013	92 340 000
166	Nicaragua	397	0.013	6 071 000
167	Rwanda	700	0.013	10 718 000
168	Mali	896	0.012	14 528 000
169	Lesotho	126	0.012	2 171 000
170	Bangladesh	7786	0.011	142 319 000
171	Côte d'Ivoire	1149	0.011	21 395 000
172	Sudan	1640	0.011	30 894 000
173	Guatemala	763	0.010	14 714 000
174	Togo	313	0.010	6 191 000
175	Madagascar	1039	0.010	20 696 000
176	Djibouti	40	0.010	818 000
177	Ethiopia	3846	0.0091	84 321 000
178	El Salvador	273	0.0088	6 200 000
179	Indonesia	9980	0.0084	237 641 000
180	Honduras	348	0.0083	8 385 072
181	Tajikistan	312	0.0080	7 800 000
182	Mauritania	128	0.0077	3 341 000
183	Yemen	895	0.0073	24 527 000
184	Dominican Republic	338	0.0072	9 379 000
185	Mozambique	841	0.0071	23 700 000
186	Sierra Leone	182	0.0067	5 400 000
187	Equatorial Guinea	22	0.0063	700 000
188	Central African Republic	155	0.0062	5 000 000

TABLE B.2 Country Population (in 2013), Number of Papers Published from 2010 to 2014 $N_{2010\text{-}2014}$, and Rank According to the Number of Papers Published Per Year from 2010 to 2014 ($\dot{N}_{2010\text{-}2014}$) Per 1000 Persons—Cont'd

Rank	Country	$N_{2010\text{-}2014}$	$\dot{N}_{2010\text{-}2014}$ per 1000	Population 2013
189	Comoros	22	0.0060	735 000
190	Niger	479	0.0059	16 275 000
191	Haiti	289	0.0057	10 085 000
192	Samoa	4	0.0043	186 000
193	Liberia	68	0.0039	3 477 000
194	Guinea	196	0.0037	10 537 000
195	Turkmenistan	79	0.0031	5 105 000
196	Eritrea	77	0.0026	6 000 000
197	Burundi	116	0.0023	10 200 000
198	Afghanistan	273	0.0022	24 485 000
199	Democratic Republic of the Congo	704	0.0021	65 966 000
200	Angola	202	0.0020	20 609 000
201	Chad	82	0.0015	11 274 000
202	Myanmar	337	0.0014	47 963 000
203	North Korea	128	0.0011	24 052 000
204	East Timor	3	0.0006	1 066 000
205	Somalia	24	0.0005	9 797 000

Bibliography

Acco, S., Williamson, D.L., van Sark, W.G.J.H.M., Sinke, W.C., van der Weg, W.F., Polman, A., Roorda, S., 1998. Nanoclustering of hydrogen in ion-implanted and plasma-grown amorphous silicon. Phys. Rev. B 58 (19), 12853-12864.

Acin-Perez, R., Enriquez, J.A., 2014. The function of the respiratory supercomplexes: the plasticity model. Biochim. Biophys. Acta 1837 (4), 444-450.

Agewall, S., Giannitsis, E., Jernberg, T., Katus, H., 2011. Troponin elevation in coronary vs. non-coronary disease. Eur. Heart J. 32 (4), 404-411.

Aken, D.C.V., Hosford, W.F., 2008. Reporting Results: A Practical Guide for Engineers and Scientists. Cambridge University Press, UK.

Akhavan, O., Bijanzad, K., Mirsepah, A., 2014. Synthesis of graphene from natural and industrial carbonaceous wastes. RSC Adv. 4 (39), 20441-20448.

Akita, T., Okumura, M., Tanaka, K., Ohkuma, K., Kohyama, M., Koyanagi, T., Date, M., Tsubota, S., Haruta, M., 2005. Transmission electron microscopy observation of the structure of TiO_2 nanotube and Au/TiO_2 nanotube catalyst. Surf. Interface Anal. 37 (2), 265-269.

Álamo, J.C.D., Jiménez, J., 2009. Estimation of turbulent convection velocities and corrections to Taylor's approximation. J. Fluid Mech. 640, 5-26.

Altschul, S.F., Gish, W., Miller, W., Myers, E.W., Lipmman, D.J., 1990. Basic local alignment search tool. J. Mol. Biol. 215 (3), 403-410.

Altschul, S.F., Madden, T.L., Schäffer, A.A., Zhang, J., Zhang, Z., Miller, W., Lipman, D.J., 1997. Gapped BLAST and PSI-BLAST: a new generation of protein database search programs. Nucleic Acids Res. 25 (17), 3389-3402.

American Association for the Advancement of Science, 2014. Science journals/AAAS authorship form and statement of conflicts of interest. URL http://stke.sciencemag.org/about/SciSignalAuthorshipform.pdf.

ARWU, 2014. Academic Ranking of World Universities 2014. URL http://www.shanghairanking.com/ARWU2014.html.

Ashburner, M., Ball, C.A., Blake, J.A., Botstein, D., Butler, H., Cherry, J.M., Davis, A.P., Dolinski, K., Dwight, S.S., Eppig, J.T., Harris, M.A., Hill, D.P., Issel-Tarver, L., Kasarskis, A., Lewis, S., Matese, J.C., Richardson, J.E., Ringwald, M., Rubin, G.M., Sherlock, G., Gene Ontology Consortium, 2000. Gene ontology: tool for the unification of biology. Nat. Genet. 25 (1), 25-29.

Ashkin, A., Dziedzic, J.M., Yamane, T., 1987. Optical trapping and manipulation of single cells using infrared laser beams. Nature 330 (6150), 769-771.

Aurore, G., Parfait, B., Fahrasmane, L., 2009. Bananas, raw materials for making processed food products. Trends Food Sci. Technol. 20 (2), 78-91.

Bang, Y.J., Cutsem, E.V., Feyereislova, A., Chung, H.C., Shen, L., Sawaki, A., Lordick, F., Ohtsu, A., Omuro, Y., Satoh, T., Aprile, G., Kulikov, E., Hill, J., Lehle, M., Ruschoff, J., Kang, Y.K., 2010. Trastuzumab in combination with chemotherapy versus chemotherapy alone for treatment of her2-positive advanced gastric or gastro-oesophageal junction cancer (toga): a phase 3, open-label, randomised controlled trial. Lancet 376 (9742), 687-697.

Baringhaus, J., Ruan, M., Edler, F., Tejeda, A., Sicot, M., Taleb-Ibrahimi, A., Li, A.P., Jiang, Z., Conrad, E.H., Berger, C., Tegenkamp, C., de Heer, W.A., 2014. Exceptional ballistic transport in epitaxial graphene nanoribbons. Nature 506 (7488), 349-354.

Barnea, A., Rubin, A., 2010. Corporate social responsibility as a conflict between shareholders. J. Bus. Ethics 97 (1), 71-86.

Bazerman, C., 1988. Shaping Written Knowledge. University of Wisconsin Press, Madison, WI.

Becke, A.D., 1988. Density-functional exchange-energy approximation with correct asymptotic behavior. Phys. Rev. A 38 (6), 3098-3100.

Becke, A.D., 1993. Density-functional thermochemistry. III. The role of exact exchange. J. Chem. Phys. 98 (7), 5648-5652.

Becke, A.D., 2014. Perspective: fifty years of density-functional theory in chemical physics. J. Chem. Phys. 140 (18), 18A301.

Bek, S., Kemler, R., 2002. Protein kinase CKII regulates the interaction of β-catenin with α-catenin and its protein stability. J. Cell Sci. 115 (24).

Benjamini, Y., Hochberg, Y., 1995. Controlling the false discovery rate—a practical and powerful approach to multiple testing. J. R. Stat. Soc. Ser. B: Stat. Methodol. 57 (1), 289-300.

Berman, H.M., Westbrook, J., Feng, Z., Gilliland, G., Bhat, T.N., Weissig, H., Shindyalov, I.N., Bourne, P.E., 2000. The Protein Data Bank. Nucleic Acids Res. 28 (1), 235-242.

Berryman, K.R., Cochran, U.A., Clark, K.J., Biasi, G.P., Langridge, R.M., Villamor, P., 2012. Major earthquakes occur regularly on an isolated plate boundary fault. Science 336 (6089), 1690-1693.

Bibes, M., Villegas, J.E., Barthelemy, A., 2011. Ultrathin oxide films and interfaces for electronics and spintronics. Adv. Phys. 60 (1), 5-84.

BIPM, 1875. International Bureau of Weights and Measures: Convention du mètre. URL http://www.bipm.org/en/worldwide-metrology/metre-convention/.

BIPM, 2006. Le Système international d'unités [updated in 2014], eighth ed. Bureau International des Poids et Mesures. URL http://www.bipm.org/utils/common/pdf/si_brochure_8_en.pdf.

BIPM, 2013. International Bureau of Weights and Measures: Mission, Role and Objectives. URL http://www.bipm.org/en/about-us/role.html.

BIPM, 2014. International Bureau of Weights and Measures: 25th Meeting of the CGPM: 18-20 November. URL http://www.bipm.org/en/cgpm-2014/.

Bitter, J.H., Seshan, K., Lercher, J.A., 1997. The state of zirconia supported platinum catalysts for CO_2/CH_4 reforming. J. Catal. 171, 279-286.

Blochl, P.E., 1994. Projector augmented-wave method. Phys. Rev. B 50 (24), 17953-17979.

Boccaletti, M., Guazzone, G., 1974. Remnant arcs and marginal basins in the Cainozoic development of the Mediterranean. Nature 252 (5478), 18-21.

Boffito, D.C., Blanco, G., Rostamizadeh, M., Patience, G.S., 2014. Biofuels Reaction-Regeneration Cycles. Lignoworks AGM, Vancouver, BC.

Boffito, D.C., Neagoe, C., Edake, M., Pastor-Ramirez, B., Patience, G.S., 2013. Biofuel synthesis in a capillary fluidized bed. Catal. Today 237, 13-17.

Bolton, R.N., Lemon, K.N., 1999. A dynamic model of customers' usage of services: usage as an antecedent and consequence of satisfaction. J. Mark. Res. 36 (2), 171-186.

Bone, R.C., Balk, R.A., Cerra, F.B., Dellinger, R.P., Fein, A.M., Knaus, W.A., Schein, R.M., Sibbald, W.J., 1992. Definitions for sepsis and organ failure and guidelines for the use of innovative therapies in sepsis. Chest 101 (6), 1644-1655.

Bornmann, L., Daniel, H.D., 2007. What do we know about the h-index? J. Am. Soc. Inf. Sci. Technol. 58 (9), 1381-1385.

Bornmann, L., de Moya Anegón, F., Leydesdorff, L., 2014. Do scientific advancements lean on the shoulders of giants? A bibliometric investigation of the Ortega hypothesis. PLoS ONE 5 (10), 1-6.

Bourassa, A.E., Robock, A., Randel, W.J., Deshler, T., Rieger, L.A., Lloyd, N.D., Llewellyn, E.J.T., Degenstein, D.A., 2012. Large volcanic aerosol load in the stratosphere linked to Asian monsoon transport. Science 337 (6090), 78-81.

Brunecky, R., Alahuhta, M., Xu, Q., Donohoe, B.S., Crowley, M.F., Kataeva, I.A., Yang, S.J., Resch, M.G., Adams, M.W.W., Lunin, V.V., Himmel, M.E., Bomble, Y.J., 2013. Revealing nature's cellulase diversity: the digestion mechanism of *Caldicellulosiruptor bescii* CelA. Science 342 (6165), 1513-1516.

Brunger, A.T., Adams, P.D., Clore, G.M., DeLano, W.L., Gros, P., Grosse-Kunstleve, R.W., Jiang, J.S., Kuszewski, J., Nilges, M., Pannu, N.S., Read, R.J., Rice, L.M., Simonson, T., Warren, G.L., 1998. Crystallography & NMR system: a new software suite for macromolecular structure determination. Acta Crystallogr. D: Biol. Crystallogr. 54 (5), 905-921.

Burger Associates Inc., 1991. The Burger Writing Courses: Desk-top Manual. Glen Mills, PA.

Calcagno, V., Demoinet, E., Gollner, K., Guidi, L., Ruths, D., de Mazancourt, C., 2012. Flows of research manuscripts among scientific journals reveal hidden submission patterns. Science 338 (6110), 1065-1069.

Canada Minister of Justice, 2012. Copyright act C-42.

Caplan, M.J., Stow, J.L., Newman, A.P., Madri, J., Anderson, H.C., Farquhar, M.G., Palade, G.E., Jamieson, J.D., 1987. Dependence on pH of polarized sorting of secreted proteins. Nature 329 (6140), 632-635.

Cavani, F., Guidetti, S., Trevisanut, C., Befi, G., Nieto, L., Soriano, M.D., Concepción, P., 2012. Tungsten-Vanadium Mixed Oxides for the Oxidehydration of Glycerol into Acrylic Acid. E-WiSPOC, Bressanone, Italy.

Chan, H.W., Wells, R.D., 1974. Structural uniqueness of lactose operator. Nature 252 (5480), 205-209.

Chang, J.S., Liao, P.H., 1999. Molecular weight control of a batch polymerization reactor: experimental study. Ind. Eng. Chem. Res. 38 (1), 144-153.

Chaput, P., Danos, V., Panangaden, P., Plotkin, G., 2014. Approximating Markov processes by averaging. J. ACM 61 (1), 5:1-5:45.

Cheng, Z., Zaki, A.A., Hui, J.Z., Muzykantov, V.R., Tsourkas, A., 2012. Multifunctional nanoparticles: cost versus benefit of adding targeting and imaging capabilities. Science 338 (6109), 903-910.

Cheong, G.L.M., Lim, K.S., Jakubowicz, A., Martens, P.J., Poole-Warren, L.A., Green, R.A., 2014. Conductive hydrogels with tailored bioactivity for implantable electrode coatings. Acta Biomater. 10 (3), 1216-1226.

Chevreul, M.E., 1855. The Principles of Harmony and Contrast of Colours, and Their Applications to the Arts, second ed. Longman, Brown, Green, and Longmans, London.

Chinchilla, R., Nájera, C., 2014. Chemicals from alkynes with palladium catalysts. Chem. Rev. 114 (3), 1783-1826.

Chiron, F.X., Patience, G.S., 2012. Kinetics of mixed copper-iron based oxygen carriers for hydrogen production by chemical looping water splitting. Int. J. Hydrogen Energy 37 (14), 10526-10538.

Choy, A., Dancourt, J., Mugo, B., O'Connor, T.J., Isberg, R.R., Melia, T.J., Roy, C.R., 2012. The *Legionella* effector RavZ inhibits host autophagy through irreversible Atg8 deconjugation. Science 338 (6110), 1072-1076.

CiteFactor, 2012. Impact Factor List. URL http://www.citefactor.org/journal-impact-factor-list-2012.html.

Cleeman, J.I., Grundy, S.M., Becker, D., Clark, L.T., Cooper, R.S., Denke, M.A., Howard, W.J., Hunninghake, D.B., Illingworth, D.R., Luepker, R.V., McBride, P., McKenney, J.M., Pasternak, R.C., Stone, N.J., Horn, L.V., Brewer, H.B., Ernst, N.D., Gordon, D., Levy, D., Rifkind, B., Rossouw, J.E., Savage, P., Haffner, S.M., Orloff, D.G., Proschan, M.A., Schwartz, J.S., Sempos, C.T., Shero, S.T., Murray, E.Z., Natl. Cholesterol Educ. Program Expe., 2001. Executive summary of the Third Report of the National Cholesterol Education Program (NCEP) expert panel on detection, evaluation, and treatment of high blood cholesterol in adults (Adult Treatment Panel III). J. Am. Med. Assoc. 285 (19), 2486-2497.

Colin Purrington, 2014. Designing conference posters. URL http://colinpurrington.com/tips/academic/posterdesign.

Coope, G.R., Lister, A.M., 1987. Late-glacial mammoth skeletons from Condover, Shropshire, England. Nature 330 (6147), 472-474.

COPE, 2008. Code of Conduct and Best Practice Guidelines for Journal Editors Committee on Publication Ethics (COPE). URL http://publicationethics.org/files/u7140/plagiarism%20A.pdf.

COPE, 2011. Code of Conduct and Best Practice Guidelines for Journal Editors Committee on Publication Ethics (COPE). URL http://publicationethics.org/files/Code%20of%20Conduct.pdf.

Coppock, B.L., Desta, S., Tezera, S., Gebru, G., 2011. Capacity building helps pastoral women transform impoverished communities in Ethiopia. Science 334 (6061), 1394-1398.

Cozzini, S., Ronchetti, M., 1996. Local icosahedral structures in binary-alloy clusters from molecular-dynamics simulation. Phys. Rev. B 53 (18), 12040-12049.

Crouse, S.F., Berent, R., von Duvillard, S.P., Schmid, P., Green, J.S., 2006. Four weeks of exercise rehabilitation improves exercise capacity and coronary risk in women with CVD. URL http://www.docstoc.com/docs/28573448/ExperimentalBiologyPoster2006-PowerPoint-PowerPoint.

Ćuk, M., Stewart, S.T., 2012. Making the moon from a fast-spinning earth: a giant impact followed by resonant despinning. Science 338 (6110), 1047-1052.

Daniel, M.C., Astur, D., 2004. Gold nanoparticles: assembly, supramolecular chemistry, quantum-size-related properties, and applications toward biology, catalysis, and nanotechnology. Chem. Rev. 104 (1), 293-346.

Davis, M., 1997. Scientific Papers and Presentations. Academic Press, Waltham, MA.

Dawood, F.S., Jain, S., Finelli, L., Shaw, M.W., Lindstrom, S., Garten, R.J., Gubareva, L.V., Xu, X., Bridges, C.B., Uyeki, T.M., 2009. Emergence of a novel swine-origin influenza a (H1N1) virus in humans novel swine-origin influenza a (H1N1) virus investigation team. N. Engl. J. Med. 360 (25), 2605-2615.

Day, R.A., Gastel, B., 2011. How to Write and Publish a Scientific Paper, seventh ed. Greenwood, Westport.

Dean, C.R., Wang, L., Maher, P., Forsythe, C., Ghahari, F., Gao, Y., Katoch, J., Ishigami, M., Moon, P., Koshino, M., Taniguchi, T., Watanabe, K., Shepard, K.L., Hone, J., Kim, P., 2013. Hofstadter's butterfly and the fractal quantum Hall effect in moire superlattices. Nature 497 (7451), 598-602.

Diwu, Z., Zimmermann, J., Meyer, T., Lown, J.W., 1994. Design, synthesis and investigation of mechanisms of action of novel protein kinase c inhibitors: perylenequinonoid pigments. Biochem. Pharmacol. 47 (2), 373-385.

Dlugan, A., 2012. What is the average speaking rate? URL http://sixminutes.dlugan.com/speaking-rate/.

Dorj, B., Won, J.E., Purevdorj, O., Patel, K.D., Kim, J.H., Lee, E.J., Kim, H.W., 2014. A novel therapeutic design of microporous-structured biopolymer scaffolds for drug loading and delivery. Acta Biomater. 10 (3), 1238-1250.

Dreyfuss, H., 1984. Symbol Sourcebook, first ed. John Wiley & Sons, New York.

Druon, J.N., Fiorentino, F., Murenu, M., Knittweis, L., Colloca, F., Osio, C., Mérigot, B., Garofalo, G., Mannini, A., Jadaud, A., Sbrana, M., Scarcella, G., Tserpes, G., Peristeraki, P., Carlucci, R., Heikkonen, J., 2015. Modelling of European hake nurseries in the Mediterranean Sea: an ecological niche approach. Prog. Oceanogr. 130, 188-204.

Dudzik, E., Norris, A.G., McGrath, R., Charlton, G., Thornton, G., Murphy, B., Turner, T.S., Norman, D., 1998. Potassium-induced removal of the $Ni(100)(2 \times 2)p_4g$-N reconstruction determined by surface X-ray diffraction. Phys. Rev. B 58 (19), 12659-12662.

Dunning, T.H., 1989. Gaussian-basis sets for use in correlated molecular calculations. 1. The atoms boron through neon and hydrogen. J. Chem. Phys. 90 (2), 1007-1023.

Dupret, F., Nicodème, P., Ryckmans, Y., Wouters, P., Crochet, M.J., 1990. Global modelling of heat transfer in crystal growth furnaces. Int. J. Heat Mass Transf. 33 (9), 1849-1871.

Edake, M., Boffito, D.C., Neagoe, C., Ramirez, B.P., Patience, G.S., 2013. Biodiesel via catalytic gas phase trans-esterification in a fluidized bed, bioenergy IV: innovations in biomass conversion for heat, power, fuels and chemicals, 2013, Otranto, Italy.

Eilola, K., Gustafsson, B.G., Kuznetsov, I., Meier, H.E.M., Neumann, T., Savchuk, O.P., 2011. Evaluation of biogeochemical cycles in an ensemble of three state-of-the-art numerical models of the Baltic Sea. J. Mar. Syst. 2011, 267-284.

Elfeiry, A.A., Garcia, L.A., 2010. Comparison of ordinary kriging, regression kriging, and cokriging techniques to estimate soil salinity using LANDSAT images. J. Irrig. Drain. Eng. 136, 355-364.

Elkins, M.H., Williams, H.L., Shreve, A.T., Neumark, D.M., 2013. Relaxation mechanism of the hydrated electron. Science 342 (6165), 1496-1499.

Elsevier, 2014a. Plagiarism-complaints. URL http://www.elsevier.com/editors/perk/plagiarism-complaints.

Elsevier, 2014b. Plagiarism-complaints. URL http://www.elsevier.com/editors/perk.

Elsevier B.V., 2014. Highlights. URL http://www.elsevier.com/journal-authors/highlights.

Emsley, P., Cowtan, K., 2004. Coot: model-building tools for molecular graphics. Acta Crystallogr. D: Biol. Crystallogr. 60 (12), 2126-2132.

Englander, K., 2014. Writing and Publishing Science Research Papers in English: A Global Perspective. Springer, Berlin.

Errami, M., Sun, Z., Long, T.C., George, A.C., Garner, H.R., 2007. Déjà vu: a study of duplicate citations in medline. Bioinformatics 24 (2), 243-249.

Errami, M., Sun, Z., Long, T.C., George, A.C., Garner, H.R., 2009. Déjà vu: a database of highly similar citations in the scientific literature. Nucleic Acids Res. 37, D921-D924.

Escandón, L.S., Ordóñez, S., Vega, A., Díez, F.V., 2005. Oxidation of methane over palladium catalysts: effect of the support. Chemosphere 58 (1), 9-17.

Fear, S., 2005. Publication quality tables in LaTeX. URL http://mirror.its.dal.ca/ctan/macros/latex/contrib/booktabs/booktabs.pdf.

Feeney, W.E., Medina, I., Somveille, M., Heinsohn, R., Hall, M.L., Mulder, R.A., Stein, J.A., Kilner, R.M., Langmore, N.E., 2013. Brood parasitism and the evolution of cooperative breeding in birds. Science 342 (6165), 1506-1508.

Friedmann-Morvinski, D., Bushong, E.A., Ke, E., Soda, Y., Marumoto, T., Singer, O., Ellisman, M.H., Verma, I.M., 2012. Dedifferentiation of neurons and astrocytes by oncogenes can induce gliomas in mice. Science 338 (6110), 1080-1084.

Frostegard, A., Baath, E., 1996. The use of phospholipid fatty acid analysis to estimate bacterial and fungal biomass in soil. Biol. Fertil. Soils 22 (1-2), 59-65.

Fujii, M.S., Zwart, S.P., 2011. The origin of OB runaway stars. Science 334 (6061), 1380-1383.

Gaese, F., Kolbeck, R., Barde, Y.A., 1994. Sensory ganglia require neurotrophin-3 early in development. Development 120 (6), 1613-1619.

Galli, F., Corbetta, M., Pirola, C., Manenti, F., 2014. Robust kinetic modelling of heterogeneously catalyzed free fatty acids esterification in (monophasic liquid)/solid packed-bed reactor: rival model discrimination. Chem. Eng. Trans. 39, 979-984.

Gallo, C., 2014. Talk Like TED. St. Martin's Griffin, New York.

García, F., Arias, J.L., Mayor, B., Pou, J., Rehman, I., Knowles, J., Best, S., León, B., Pérez-Amor, M., Bonfield, W., 1998. Effect of heat treatment on pulsed laser deposited amorphous calcium phosphate coatings. J. Biomed. Mater. Res. 43 (1), 69-76.

Geim, A.K., Novoselov, K.S., 2007. The rise of graphene. Nat. Mater. 6 (3), 183-191.

Gelb, M.J., 1988. Present Yourself!. Jalmar Press, Torrance, CA.

Ghaznavi, T., Boffito, D.C., Neagoe, C., Patience, G.S., 2014. Xylose Oxidation in a Capillary Fluidized Bed. Lignoworks AGM, Vancouver, BC.

Gingras, Y., Larivière, V., Macaluso, B., Robit, J.P., 2008. The effects of aging of scientists on their publication and citation patterns. URL http://arxiv.org/ftp/arxiv/papers/0810/0810.4292.pdf.

Gladwell, M., 2008. Outliers. Little, Brown and Company, New York.

Godini, H.R., Jašo, S., Xiao, S., Arellano-Garcia, H., Omidkhah, M., Wozny, G., 2012. Methane oxidative coupling: synthesis of membrane reactor networks. Ind. Eng. Chem. Res. 51 (22), 7747-7761.

Gómez-Estaca, J., López-de-Dicastillo, C., Hernández-Muñoz, P., Catalá, R., Gavara, R., 2014. Advances in antioxidant active food packaging. Trends Food Sci. Technol. 35 (1), 42-51.

González-Pereira, B., Guerrero-Bote, V.P., Moya-Anegón, F., 2013. The SJR indicator: a new indicator of journals' scientific prestige.

Goodman, P.A., Li, H., Gao, Y., Lu, Y.F., Stenger-Smith, J.D., Redepenning, J., 2013. Preparation and characterization of high surface area, high porosity carbon monoliths from pyrolyzed bovine bone and their performance as supercapacitor electrodes. Carbon 55, 291-298.

Google Scholar, 2014. URL http://scholar.google.ca/citations?view_op=top_venues&hl=en.

Gotoh, K., Finney, J.L., 1974. Statistical geometrical approach to random packing density of equal spheres. Nature 252 (5480), 202-205.

Green, D.J., Alemseged, Z., 2012. *Australopithecus afarensis* scapular ontongen, function and the role of climbing in human evolution. Science 338 (6106), 514-517.

Grefenstette, B.W., Harrison, F.A., Boggs, S.E., Reynolds, S.P., Fryer, C.L., Madsen, K.K., Wik, D.R., Zoglauer, A., Ellinger, C.I., Alexander, D.M., An, H., Barret, D., Christensen, F.E., Craig, W.W., Forster, K., Giommi, P., Hailey, C.J., Hornstrup, A., Kaspi, V.M., Kitaguchi, T., Koglin, J.E., Mao, P.H., Miyasaka, H., Mori, K., Perri, M., Pivovaroff, M.J., Puccetti, S., Rana, V., Stern, D., Westergaard, N.J., Zhang, W.W., 2014. Asymmetries in core-collapse supernovae from maps of radioactive ^{44}Ti in CassiopeiaA. Nature 506 (7488), 339-342.

Grundmeier, G., Schmidt, W., Stratmann, M., 2000. Corrosion protection by organic coatings: electrochemical mechanism and novel methods of investigation. Electrochim. Acta 45 (15), 2515-2533.

Gupta, H.M., Campanha, J.R., Pesce, R.A.G., 2005. Power-law distributions for the citation index of scientific publications and scientists. Braz. J. Phys. 35 (4a), 981-986.

Hammarlund, P., Martinez, A.J., Bajwa, A.A., Hill, D.L., Hallnor, E., Jiang, H., Dixon, M., Derr, M., Hunsaker, M., Kumar, R., Osborne, R.B., Rajwar, R., Singhal, R., D'Sa, R., Chappell, R., Kaushik, S., Chennupaty, S., Jourdan, S., Gunther, S., Piazza, T., Burton, T., 2014. Haswell: the fourth-generation Intel Core Processor. IEEE Micro 34 (2), 6-20.

Hanahan, D., Weinberg, R.A., 2000. The hallmarks of cancer. Cell 100 (1), 57-70.

Harzing, A.-W., 2010. Citation analysis across disciplines: the impact of different data sources and citation metrics. URL http://www.harzing.com/data_metrics_comparison.htm.

Hatzivassiloglou, V., McKeown, K.R., 1997. Predicting the semantic orientation of adjectives. In: Proceedings of the 35th Annual Meeting of the Association for Computational Linguistics and Eighth Conference of the European Chapter of the Association for Computational Linguistics, pp. 174-181.

Hayes, R.E., Kolaczkowski, S.T., Li, P.K.C., Awdry, S., 2001. The palladium catalysed oxidation of methane: reaction kinetics and the effect of diffusion barriers. Chem. Eng. Sci. 56 (16), 4815-4835.

Heldoorn, M., 2007. The SI units package—consistent application of SI units. URL http://texdoc.net/texmf-dist/doc/latex/SIunits/SIunits.pdf.

Hettiarachchi, D.J., 2014. I can see something. URL http://www.businessinsider.com/toastmasters-public-speaking-champion-dananjaya-hettiarachchi-2014-9.

Hirsch, J.E., 2005. An index to quantify and individual's scientific research output. Proc. Natl. Acad. Sci. U. S. A. 102 (46), 16569-16572.

Hobaiter, C., Byrn, R.W., 2011. The gestural repertoire of the wild chimpanzee. Anim. Cogn. 14, 745-767.

Hohenberg, P., Kohn, W., 1964. Inhomogeneous electron gas. Phys. Rev. 136 (3B), B864-B871.

Horwitz, E.K., Horwitz, M.B., Cope, J., 1986. Foreign language classroom anxiety. Mod. Lang. J. 70 (2), 125-132.

Hu, L., Bentler, P.M., 1999. Cutoff criteria for fit indexes in covariance structure analysis: conventional criteria versus new alternatives. Struct. Equ. Model.: Multidisciplinary J. 6 (1), 1-55.

Hu, Q., Yuan, Q., Deng, Y., Ling, Y., Tang, H., 2015. Synthesis and solid-state properties of thermotropic liquid crystalline polypeptide bearing imidazolium and p-tolyl groups. Eur. Polym. J. 63, 74-79.

Hudson, J.B., Imperial, V., Haugland, R.P., Diwu, Z., 1997. Antiviral activities of photoactive perylenequinones. Photochem. Photobiol. 65 (2), 352-354.

Huelsenbeck, J.P., Ronquist, F., 2001. MRBAYES: Bayesian inference of phylogenetic trees. Bioinformatics 17 (8), 754-755.

Humphrey, W., Dalke, A., Schulten, K., 1996. VMD: visual molecular dynamics. J. Mol. Graph. Model. 14 (1), 33-38.

Hurricane, O.A., Callahan, D.A., Casey, D.T., Celliers, P.M., Cerjan, C., Dewald, E.L., Dittrich, T.R., Döppner, T., Hinkel, D.E., Hopkins, L.F.B., Kline, J.L., Pape, S.L., Ma, T., MacPhee, A.G., Milovich, J.L., Pak, A., Park, H.S., Patel, P.K., Remington, B.A., Salmonson, J.D., Springer, P.T., Tommasini, R., 2014. Fuel gain exceeding unity in an inertially confined fusion implosion. Nature 506 (7488), 343-348.

Hyland, K., 2005. Metadiscourse. Continuum, London & New York.

Iijima, S., 1991. Helical microtubules of graphitic carbon. Nature 354 (6348), 56-58.

Inskeep, W.P., Macur, R.E., Hamamura, N., Warelow, T.P., Ward, S.A., Santini, J.M., 2007. Detection, diversity and expression of aerobic bacterial arsenite oxidase genes. Environ. Microbiol. 9 (4), 934-943.

International Committee of Medical Journal Editors, 2014. Defining the role of authors and contributors. URL http://www.icmje.org/recommendations/browse/roles-and-responsibilities/defining-the-role-of-authors-and-contributors.html.

Ioannidis, J.P.A., Boyack, K.W., Klavans, R., 2014. Estimates of the continuously publishing core in the scientific workforce. PLOS ONE 9 (7), 1-10.

Jablan, M., Buljan, H., Soljačić, M., 2009. Plasmonics in graphene at infrared frequencies. Phys. Rev. B 80 (245435), 1-7.

Jahn, M., Rogers, M.J., Söll, D., 1991. Anticodon and acceptor stem nucleotides in tRNA^{Gln} are major recognition elements for E. coli glutaminyl-tRNA synthetase. Nature 352 (6332), 258-260.

Jankowski, P., McKellar, A.R.W., Szalewicz, K., 2012. Theory untangles the high-resolution infrared spectrum of the *ortho*-h_2-co van der Waals complex. Science 336 (6085), 1147-1150.

JCGM, 2008. Joint Committee for Guides in Metrology. Evaluation of Measurement Data—Guide to the Expression of Uncertainty in Measurement. URL www.bipm.org.

Jin, J., 2014. Stroke risk may be increased after shingles episodes. J. Am. Med. Assoc. 311 (17), 1718.

Jinek, M., Chylinski, K., Fonfara, I., Hauer, M., Doudna, J.A., Charpentier, E., 2012. A programmable dual-RNA-guided DNA endonuclease in adaptive bacterial immunity. Science 337 (6096), 816-821.

Jones, T.A., Zou, J.Y., Cowan, S.W., Kjeldgaard, M., 1991. Improved methods for building protein models in electron density maps and the location of errors in these models. Acta Crystallogr. A 47 (2), 110-119.

Joughin, I., Alley, R.B., Holland, D.M., 2012. Ice-sheet response to oceanic forcing. Science 338 (6111), 1172-1176.

Julien, J.P., Cupo, A., Sok, D., Stanfield, R.L., Lyumkis, D., Deller, M.C., Klasse, P.J., Burton, D.R., Sanders, R.W., Moore, J.P., Ward, A.B., Wilson, I.A., 2013. Crystal structure of a soluble cleaved HIV-1 envelope trimer. Science 342 (6165), 1477-1483.

Kaarsholm, M., Nerlov, J., Cenni, R., Chaouki, J., Patience, G.S., 2006. Activity Profile of Modified ZSM-5 for Methanol-to-Olefins. ISCRE19, Potsdam, Germany.

Kalnay, E., Kanamitsu, M., Kistler, R., Collins, W., Deaven, D., Gandin, L., Iredell, M., Saha, S., White, G., Woollen, J., Zhu, Y., Chelliah, M., Ebisuzaki, W., Higgins, W., Janowiak, J., Mo, K.C., Ropelewski, C., Wang, J., Leetmaa, A., Reynolds, R., Jenne, R., Joseph, D., 1996. The NCEP/NCAR 40-year reanalysis project. Bull. Am. Meteorol. Soc. 77 (3), 437-471.

Kamat, P.V., 2015. ACS Webinar series: mastering the art of scientific publication—preparing your manuscript: Part 1. URL http://event.on24.com/eventRegistration/EventLobbyServlet?target=lobby.jsp&eventid=919992&sessionid=1&key=F4BE615F725E6954C134C751EF165C61&eventuserid=110790824.

Kamat, P.V., Hartland, G.V., Schatz, G.C., 2014a. Graphical excellence. J. Phys. Chem. Lett. 5 (12), 2118-2120.

Kamat, P.V., Schaltz, G.C., 2013. Cite with a sight. J. Phys. Chem. Lett. 4 (9), 1578-1581.

Kamat, P.V., Schaltz, G.C., 2014. How to make your next paper scientifically effective. J. Phys. Chem. Lett. 5 (7), 1241-1242.

Kamat, P.V., Sholes, G., Prezhdo, O., Zaera, F., Zwier, T., Schatz, G.C., 2014b. Overcoming the myths of the review process and getting your paper ready for publication. J. Phys. Chem. Lett. 5 (5), 896-899.

Karirekinyana, G., Patience, G.S., 2010. Anti-Malaria and Sustainable Development in Burundi, International Conference on Malaria and Sustainable Development 2010, Bujumbura, Burundi.

Ke, Y., Ong, L.L., Shih, W.M., Yin, P., 2012. Three-dimensional structures self-assembled from DNA bricks. Science 338 (6111), 1177-1183.

Keller, K.E., Tan, I.S., Lee, Y.S., 2012. SAICAR stimulates pyruvate kinase isoform M2 and promotes cancer cell survival in glucose-limited conditions. Science 338 (6110), 1069-1072.

Kim, H.J., Kim, N.C., Wang, Y.D., Scarborough, E.A., Moore, J., Diaz, Z., MacLea, K.S., Freibaum, B., Li, S.Q., Molliex, A., Kanagaraj, A.P., Carter, R., Boylan, K.B., Wojtas, A.M., Rademakers, R., Pinku, J.L., Greenberg, S.A., Trojanowski, J.Q., Traynor, B.J., Smith, B.N., Topp, S., Gkazi, A.S., Miller, J., Shaw, C.E., Kottlors, M., Kirschner, J., Pestronk, A., Li, Y.R., Ford, A.F., Gitler, A.D., Benatar, M., King, O.D., Kimonis, V.E., Ross, E.D., Weihl, C.C., Shorter, J., Taylor, J.P., 2013. Mutations in prion-like domains in hnRNPA2B1 and hnRNPA1 cause multisystem proteinopathy and ALS. Nature 495 (7442), 467.

Kirkman, J., 1971. What is Good Style for Engineering Writing?. The Institution of Chemical Engineers, Rugby, UK.

Kirste, M., Wang, X., Schewe, H.C., Meijer, G., Liu, K., van der Avoird, A., Janssen, L.M.C., Gubbels, K.B., Groenenboom, G.C., van de Meerakker, S.Y.T., 2012. Quantum-state resolved bimolecular collision of velocity-controlled OH with NO radicals. Science 338 (6110), 1060-1063.

Kohler, E., 2014. LaTeX usage notes. URL http://www.read.seas.harvard.edu/kohler/latex.html.

Kohn, W., Sham, L.J., 1965. Self-consistent equations including exchange and correlation effects. Phys. Rev. 140 (4A), A1133-A1138.

Kraulis, P.J., 1991. Molscript—a program to produce both detailed and schematic plots of protein structures. J. Appl. Crystallogr. 24 (5), 946-950.

Kresge, C.T., Leonowicz, M.E., Roth, W.J., Vartuli, J.C., Beck, J.S., 1992. Ordered mesoporous molecular-sieves synthesized by a liquid-crystal template mechanism. Nature 359 (6397), 710-712.

Kresse, G., Furthmüller, J., 1996a. Efficiency of ab-initio total energy calculations for metals and semiconductors using a plane-wave basis set. Comput. Mater. Sci. 6 (1), 15-50.

Kresse, G., Furthmüller, J., 1996b. Efficient iterative schemes for *ab initio* total-energy calculations using a plane-wave basis set. Phys. Rev. B 54 (16), 11169-11186.

Kresse, G., Hafner, J., 1993. Ab initio molecular-dynamics for liquid-metals. Phys. Rev. B 47 (1), 558-561.

Kresse, G., Joubert, D., 1999. From ultrasoft pseudopotentials to the projector augmented-wave method. Phys. Rev. B 59 (3), 1758-1775.

Krzywinski, M., 2013. Points of significance: axes, ticks and grids. Nat. Methods 10 (3), 183.

Krzywinski, M., Altman, N., 2013a. Points of significance: error bars. Nat. Methods 10 (10), 921-922.

Krzywinski, M., Altman, N., 2013b. Points of significance: power and sample size. Nat. Methods 10 (12), 1139-1140.

Krzywinski, M., Wong, B., 2013. Points of significance: plotting symbols. Nat. Methods 10 (6), 451.

Kuehn, B.M., 2014. Nationwide IV fluid shortage threatens care. J. Am. Med. Assoc. 311 (18), 1843-1844.

Kuila, T., Bose, S., Mishra, A.K., Khanra, P., Kim, N.H., Lee, J.H., 2012. Effect of functionalized graphene on the physical properties of linear low density polyethylene nanocomposites. Polym. Test. 31, 31-38.

Kusnetsov, V.V., Podlovchenko, B.I., Shakurov, R.I., Kavyrshina, K.V., Lyahenko, S.E., 2014. nPt^0 (H$_{x-2n}$ MoO$_3$) as a promising catalyst for the oxidation of methanol. synthesis and electrocatalytic properties. Int. J. Hydrogen Energy 39 (2), 829-836.

Lai, F., Orom, U.A., Cesaroni, M., Beringer, M., Taatjes, D.J., Blobel, G.A., Shiekhattar, R., 2013. Activating RNAs associate with mediator to enhance chromatin architecture and transcription. Nature 494 (7438), 497-501.

Lai, K., Feldman, M., Stoica, I., Chuang, J., 2003. Incentives for cooperation in peer-to-peer networks. In: Workshop on Economics of Peer-to-Peer Systems, pp. 1243-1248.

Lai, Y., Liang, X., Yang, S., Liu, P., Zeng, Y., Hu, C., 2014. Raman and FTIR spectra of CeO$_2$ and Gd$_2$O$_3$ in iron phosphate glasses. J. Alloys Compd. 617, 597-601.

Lambin, E.F., Meyfroidt, P., 2010. Land use transitions: socio-ecological feedback versus socio-economic change. Land Use Policy 27 (2), 108-118.

Lan, J., Zhao, Y., Dong, F., Yan, Z., Zheng, W., Fan, J., Sun, G., 2015. Meta-analysis of the effect and safety of berberine in the treatment of type 2 diabetes mellitus, hyperlipemia and hypertension. J. Ethnopharmacol. 161, 69-81.

Lander, E.S., Int Human Genome Sequencing Consortium, Linton, L.M., Birren, B., Nusbaum, C., Zody, M.C., Baldwin, J., Devon, K., Dewar, K., Doyle, M., FitzHugh, W., Funke, R., Gage, D., Harris, K., Heaford, A., Howland, J., Kann, L., Lehoczky, J., LeVine, R., McEwan, P., McKernan, K., Meldrim, J., Mesirov, J.P., Miranda, C., Morris, W., Naylor, J., Raymond, C., Rosetti, M., Santos, R., Sheridan, A., Sougnez, C., Stange-Thomann, N., Stojanovic, N., Subramanian, A., Wyman, D., Rogers, J., Sulston, J., Ainscough, R., Beck, S., Bentley, D., Burton, J., Clee, C., Carter, N., Coulson, A., Deadman, R., Deloukas, P., Dunham, A., Dunham, I., Durbin, R., French, L., Grafham, D., Gregory, S., Hubbard, T., Humphray, S., Hunt, A., Jones, M., Lloyd, C., McMurray, A., Matthews, L., Mercer, S., Milne, S., Mullikin, J.C., Mungall, A., Plumb, R., Ross, M., Shownkeen, R., Sims, S., Waterston, R.H., Wilson, R.K., Hillier, L.W., McPherson, J.D., Marra, M.A., Mardis, E.R., Fulton, L.A., Chinwalla, A.T., Pepin, K.H., Gish, W.R., Chissoe, S.L., Wendl, M.C., Delehaunty, K.D., Miner, T.L., Delehaunty, A., Kramer, J.B., Cook, L.L., Fulton, R.S., Johnson, D.L., Minx, P.J., Clifton, S.W., Hawkins, T., Branscomb, E., Predki, P., Richardson, P., Wenning, S., Slezak, T., Doggett, N., Cheng, J.F., Olsen, A., Lucas, S., Elkin, C., Uberbacher, E., Frazier, M., Gibbs, R.A., Muzny, D.M., Scherer, S.E., Bouck, J.B., Sodergren, E.J., Worley, K.C., Rives, C.M., Gorrell, J.H., Metzker, M.L., Naylor, S.L., Kucherlapati, R.S., Nelson, D.L., Weinstock, G.M., Sakaki, Y., Fujiyama, A., Hattori, M., Yada, T., Toyoda, A., Itoh, T., Kawagoe, C., Watanabe, H., Totoki, Y., Taylor, T., Weissenbach, J., Heilig, R., Saurin, W., Artiguenave, F., Brottier, P., Bruls, T., Pelletier, E., Robert, C., Wincker, P., Rosenthal, A., Platzer, M., Nyakatura, G., Taudien, S., Rump, A., Yang, H.M., Yu, J., Wang, J., Huang, G.Y., Gu, J., Hood, L., Rowen, L., Madan, A., Qin, S.Z., Davis, R.W., Federspiel, N.A., Abola, A.P., Proctor, M.J., Myers, R.M., Schmutz, J., Dickson, M., Grimwood, J., Cox, D.R., Olson, M.V., Kaul, R., Raymond, C., Shimizu, N., 2001. Initial sequencing and analysis of the human genome. Nature 409 (6822), 860-921.

Langecker, M., Arnaut, V., Martin, T.G., List, J., Renner, S., Mayer, M., Dietz, H., Simmel, F.C., 2012. Synthetic lipid membrane channels formed by designed DNA nanostructures. Science 338 (6109), 932-936.

Laskowski, R.A., Macarthur, M.W., Moss, D.S., Thornton, J.M., 1993. Procheck—a program to check the stereochemical quality of protein structures. J. Appl. Crystallogr. 26 (2), 283-291.

Lee, C., Yang, W., Parr, R.G., 1988. Development of the Colle-Salvetti correlation-energy formula into a functional of the electron density. Phys. Rev. B 37 (2), 785-789.

Leucht, S., Corves, C., Arbter, D., Engel, R.R., Li, C., Davis, J.M., 2009. Second-generation versus first-generation antipsychotic drugs for schizophrenia: a meta-analysis. Lancet 373 (9657), 31-41.

Lewandowsky, S., Spence, I., 1989. Discriminating strata in scatterplots. J. Am. Stat. Assoc. 84 (407), 682-688.

Li, R., Wu, T., Zeng, L., Xu, J., Liu, P., 2014. Effect of thermotropic liquid crystalline poly(ether ketone)arylates on the processability and properties of poly(ether ether ketone)s fibers. J. Appl. Polym. Sci. 131 (16), 40595:1-40595:9.

Lin, L., Comita, L.S., Zheng, Z., Cao, M., 2012. Seasonal differentiation in density-dependent seedling survival in a tropical rain forest. J. Ecol. 100 (4), 905-914.

Linster, M., van Boheemen, S., de Graaf, M., Schrauwen, E.J.A., Lexmond, P., Manz, B., Bestebroer, T.M., Baumann, J., van Riel, D., Rimmelzwaan, G.F., Osterhaus, A.D.M.E., Matrosovich, M., Fouchier, R.A.M., Herfst, S., 2014. Identification, characterization, and natural selection of mutations driving airborne transmission of A/H5N1 virus. Cell 157 (2), 329-339.

Liu, T., Liu, Z., Song, C., Hu, Y., Han, Z., She, J., Fan, F., Wang, J., Jin, C., Chang, J., Zhou, J.M., Chai, J., 2012. Chitin-induced dimerization activates a plant immune receptor. Science 336 (6085), 1160-1164.

Liu, Y., K. Ai, L.L., 2014. Polydopamine and its derivative materials: synthesis and promising applications in energy, environmental, and biomedical fields. Chem. Rev. 114 (9), 5057-5115.

Livak, K.J., Schmittgen, T.D., 2001. Analysis of relative gene expression data using real-time quantitative PCR and the $2^{-\Delta\Delta C_T}$ method. Methods 25 (4), 402-408.

Lowenstern, J.B., Evans, W.C., Bergfeld, D., Hunt, A.G., 2014. Prodigious degassing of a billion years of accumulated radiogenic helium at Yellowstone. Nature 506 (7488), 355-358.

Lowry, O.H., Rosebrough, N.J., Farr, A.L., Randall, R.J., 1951. Protein measurement with the folin phenol reagent. J. Biol. Chem. 193 (1), 265-275.

Maggioni, G., Castagliuolo, I., 2014. An innovative pathway towards nano-sized parylene-based materials prepared by plasma-based deposition: example of application to the production of antimicrobial materials. Plasma Process. Polym. 11 (5), 489-495.

Mäki-Arvela, P., Simakova, I.L., Salmi, T., Murzin, D.Y., 2014. Production of lactic acid/lactates from biomass and their catalytic transformations to commodities. Chem. Rev. 114 (3), 1909-1971.

Man, H.C., Zhang, S., Cheng, F.T., Yue, T.M., 2002. In situ synthesis of tic reinforced surface MMC on AL6061 by laser surface alloying. Scr. Mater. 46 (3), 229-234.

Marcus, A., Oransky, I., 2011. Banned. Lab Times 6.

Martinez-Rubio, M.I., Ireland, T.G., Fern, G.R., Silver, J., Snowden, M.J., 2000. A new application for microgels: novel method for the synthesis of spherical particles of the Y_2O_3:EUV phosphor using a copolymer microgel of NIPAM and acrylic acid. Langmuir 17 (22), 7145-7149.

McGuire, J.A., Sykora, M., Joo, J., Pietryga, J.M., Klimov, V.I., 2010. Apparent versus true carrier multiplication yields in semiconductor nanocrystals. Nano Lett. 10 (6), 2049-2057.

Merriam-Webster, 2014. URL http://www.merriam-webster.com/.

Michaelides, S., Levizzani, V., Anagnostou, E., Bauer, P., Kasparis, T., Lane, J.E., 2009. Precipitation: measurement, remote sensing, climatology and modeling. Atmos. Res. 94 (4), 512-533.

Mike, M., 2014. Cancer care shows signs of strain as patients live longer. J. Am. Med. Assoc. 311 (17), 1717-1718.

Moncada, S., Palmer, R.M.J., Higgs, E.A., 1991. Nitric-oxide—physiology, pathophysiology, and pharmacology. Pharmacol. Rev. 43 (2), 109-142.

Montolio-Rodriguez, D., Linke, D., Linke, P., 2007. Systematic identification of optimal process designs for the production of acetic acid via ethane oxidation. Chem. Eng. Sci. 62 (20), 5602-5608.

Myers, J.H., Savoie, A., van Randen, E., 1998. Eradication and pest management. Annu. Rev. Entomol. 43 (1), 471-491.

Nature, 2014. Reporting checklist for life sciences articles. URL http://www.nature.com/authors/policies/checklist.pdf.

Naturephotonics, E., 2009. Combating plagiarism. Nat. Photonics 3, 237.

Novoselov, K.S., Geim, A.K., Morozov, S.V., Jiang, D., Zhang, Y., Dubonos, S.V., Grigorieva, I.V., Firsov, A.A., 2004. Electric field effect in atomically thin carbon films. Science 306 (5696), 666-669.

NSERC, 2014. Policy and guidelines on the assessment of contributions to research and training. URL http://www.nserc-crsng.gc.ca/NSERC-CRSNG/Policies-Politiques/assesscontrib-evalcontrib_eng.asp.

O'Connor, M., 1991. Writing Successfully in Science. E & FN SPON, London.

Oliveira, C.A.B., Gehman, V., Goldschmidt, A., Nygren, D., Renner, J., 2013. First results in pure high pressure Xe towards the study of Xe + TMA mixtures with dark matter directionality sensitivity and supra-intrinsic energy resolution for $0\nu\beta\beta$ decay searches. URL http://carlosoliveiraresearch.com/images/posters/TAUP2013_poster.jpg.

O'Regan, B., Gratzel, M., 1991. A low-cost, high-efficiency solar-cell based on dye-sensitized colloidal TiO_2 films. Nature 353 (6346), 737-740.

Otwinowski, Z., Minor, W., 1997. Processing of X-ray diffraction data collected in oscillation mode. Methods Enzymol. 276, 307-326.

Paglia, E.E., Valentine, W.N., 1967. Studies on the quantitative and qualitative characterization of erythrocyte glutathione peroxidase. J. Lab. Clin. Med. 70 (1), 158-169.

Pan, J.W., Bouwmeester, D., Weinfurter, H., Zeilinger, A., 1998. Experimental entanglement swapping: entangling photons that never interacted. Phys. Rev. Lett. 80 (18), 3891-3894.

Patel, M.R., Mahaffey, K.W., Garg, J., Pan, G., Singer, D.E., Hacke, W., Breithardt, G., Halperin, J.L., Hankey, G.J., Piccini, J.P., Becker, R.C., Nessel, C.C., Paolini, J.F., Berkowitz, S.D., Fox, K.A.A., Califf, R.M., 2011. Rivaroxaban versus warfarin in nonvalvular atrial fibrillation. N. Engl. J. Med. 365 (10), 883-891.

Patience, G.S., 2013. Experimental Methods and Instrumentation for Chemical Engineers, first ed. Elsevier B.V., Amsterdam, Netherlands.

Patience, G.S., Bockrath, R.E., 2010. Butane oxidation process development in a circulating fluidized bed. Appl. Catal. A: Gen. 376, 4-12.

Patience, G.S., Boffito, D.C., Patience, P.A., 2013. Writing a scientific paper: from clutter to clarity. URL http://www.elsevier.com/__data/assets/pdf_file/0008/145943/2014-01-15-Manuscript-preparation.pdf.

Patrick, J.S., Siek, K., Binkley, J., 2013. Characterization of nutrients and actives in herbal supplements and nutriceuticals using UHPLC-reflecting time-of-flight mass spectrometry—comparison of suppliers and dosage forms of green tea, ginger, and acai berry extracts. URL http://www.eposters.net/pdfs/characterization-of-nutrients-and-actives-in-herbal-supplements-and-nutriceuticals-using-uhplc.pdf.

Pearson, C., 2014. How to tune typography based on characters per line. URL http://www.pearsonified.com/2012/01/characters-per-line.php.

Pehnt, M., Henkel, J., 2009. Life cycle assessment of carbon dioxide capture and storage from lignite power plants. Int. J. Greenhouse Gas Control 3 (1), 49-66.

Peng, Z., Freunberger, S.A., Chen, Y., Bruce, P.G., 2012. A reversible and higher-rate $Li-O_2$ battery. Science 337 (6094), 563-566.

Peramo, A., Marcelo, C.L., 2010. Bioengineering the skin-implant interface: the use of regenerative therapies in implanted devices. Ann. Biomed. Eng. 38 (6), 2013-2031.

Perdew, J.P., Burke, K., Ernzerhof, M., 1996. Generalized gradient approximation made simple. Phys. Rev. Lett. 77 (18), 3865-3868.

Perdew, J.P., Chevary, J.A., Vosko, S.H., Jackson, K.A., Pederson, M.R., Singh, D.J., Fiolhais, C., 1992. Atoms, molecules, solids, and surfaces—applications of the generalized gradient approximation for exchange and correlation. Phys. Rev. B 46 (11), 6671-6687.

Perdew, J.P., Wang, Y., 1992. Accurate and simple analytic representation of the electron-gas correlation-energy. Phys. Rev. B 45 (23), 13244-13249.

Perreault, P., Stainton, H., Yazdanpanah, M.M., Patience, G.S., 2012. Kinetic Modelling of Ilmenite Reduction by CO, H_2 and CH_4, 2nd International Conference on Chemical Looping 2012, Darmstadt, Germany.

Perret, N., Alexander, A.M., Hunter, S.M., Chung, P., Hargreaves, J.S.J., Howe, R.F., Keane, M.A., 2014a. Synthesis, characterisation and hydrogenation performance of ternary nitride catalysts. Appl. Catal. A: Gen. 468, 128-137.

Perret, N., Wang, X., Delgado, J.J., Blanco, G., Chen, X., Olmos, C.M., Bernal, S., Keane, M.A., 2014b. Selective hydrogenation of benzoic acid over Au supported on CeO_2 and $Ce_0 \cdot 62Zr_0 \cdot 38O_2$: formation of benzyl alcohol. J. Catal. 317, 114-125.

Pershin, Y.V., Ventra, M.D., 2011. Memory effects in complex materials and nanoscale systems. Adv. Phys. 60 (2), 145-227.

Picard, M., Burelle, Y., 2012. Mitochondria: starving to reach quorum? Insight into the physiological purpose of mitochondrial fusion. BioEssays 34 (4), 272-274.

Piehler, D., 2009. Legal and practical pitfalls in making use of patents. Nature 462.

Pinker, S., 2014. The Sense of Style: The Thinking Person's Guide to Writing in the 21st Century. Penguin, New York, NY.

Pirola, C., Bianchi, C.L.M., Fronzo, A.D., Boffito, D.C., Michele, A.D., Patience, G.S., 2013. Ultrasound and microwave assisted preparation of high Fe loaded supported catalysts for biosyngas Fischer-Tropsch conversion. In: ACS National Meeting & Exposition.

Pirola, C., Galli, F., Comazzi, A., Manenti, F., Bianchi, C.L., 2014. Preservation of carotenes in the deacidification of crude palm oil. RSC Adv. 4, 46922-46925.

Polytechnique, E., 2014. Règlements et sanctions. URL http://www.polymtl.ca/etudes/ppp/reglement/index.php.

Posada, D., Crandall, K.A., 1998. Modeltest: testing the model of DNA substitution. Bioinformatics 14 (9), 817-818.

Prabhanjan, D.G., Rennie, T.J., Raghavan, G.S.V., 2004. Natural convection heat transfer from helical coiled tubes. Int. J. Therm. Sci. 43 (4), 359-365.

Preston-Thomas, H., 1990. The International Temperature Scale of 1990 (ITS-90). Metrologia 27 (1), 3-10.

Prusiner, S.B., 1998. Prions. Proc. Natl. Acad. Sci. U. S. A. 95 (23), 13363-13383.

Purcaro, G., Moret, S., Conte, L.S., 2009. Optimisation of microwave assisted extraction (MAE) for polycyclic aromatic hydrocarbon (PAH) determination in smoked meat. Meat Sci. 81 (1), 275-290.

Rabinowitz, H., Vogel, S. (Eds.) 2009. The Manual of Scientific Style: A Guide for Authors, Editors, and Researchers, first ed. Academic Press, Waltham, MA.

Raymond, B., West, S.A., Griffin, A.S., Bonsall, M.B., 2012. The dynamics of cooperative bacterial virulence in the field. Science 337 (6090), 85-88.

Rehm, J., Mathers, C., Popova, S., Thavorncharoensap, M., Teerawattananon, Y., Patra, J., 2009. Global burden of disease and injury and economic cost attributable to alcohol use and alcohol-use disorders. Lancet 373, 2223-2233.

Ren, Z.A., Lu, W., Yang, J., Yi, W., Shen, X.L., Li, Z.C., Che, G.C., Dong, X.L., Sun, L.L., Zhou, F., Zhao, Z.X., 2008. Superconductivity at 55 k in iron-based F-doped layered quaternary compound SM [O1-xFx] FeAs, arxiv preprint arxiv:0804.2053.

Rezende, G., 2013. Chromatic wheel 2. URL https://openclipart.org/detail/181642/chromatic-wheel-2-by-gustavorezende-181642.

Rice, W.R., 1989. Analyzing tables of statistical tests. Evolution 43 (1), 223-225.

Rinke, C., Schwientek, P., Sczyrba, A., Ivanova, N.N., Anderson, I.J., Cheng, J.F., Darling, A., Malfatti, S., Swan, B.K., Gies, E.A., Dodsworth, J.A., Hedlund, B.P., Tsiamis, G., Sievert, S.M., Liu, W.T., Eisen, J.A., Hallam, S.J., Kyrpides, N.C., Stepanauskas, R., Rubin, E.M., Hugenholtz, P., Woyke, T., 2013. Insights into the phylogeny and coding potential of microbial dark matter. Nature 499 (7459), 431-437.

Ronquist, F., Huelsenbeck, J.P., 2003. MRBAYES 3: Bayesian phylogenetic inference under mixed models. Bioinformatics 19 (12), 1572-1574.

Rosner, J.L., 1990. Reflections of science as a product. Nature 345 (6271), 108.

Ross, R., 1999. Mechanisms of disease—atherosclerosis—an inflammatory disease. N. Engl. J. Med. 340 (2), 115-126.

Royal Society of Chemistry, 2014. Ethical guidelines and conflict of interest. URL http://www.rsc. org/Publishing/Journals/guidelines/EthicalGuidelines/EthicalGuidelinesandConflictofInterest/ sect3.asp.

Salazar, R.F., Dotson, N.M., Bressler, S.L., Gray, C.M., 2012. Content-specific fronto-parietal synchronization during visual working memory. Science 338 (6110), 1097-1110.

Schieffelin, J.S., Shaffer, J.G., Goba, A., Gbakie, M., Gire, S.K., Colubri, A., Sealfon, R.S.G., Kanneh, L., Moigboi*, A., Momoh, M., Fullah*, M., Moses, L.M., Brown, B.L., Andersen, K.G., Winnicki, S., Schaffner, S.F., Park, D.J., Yozwiak, N.L., Jiang, P.P., Kargbo, D., Jalloh, S., Fonnie*, M., Sinnah*, V., French, I., Kovoma*, A., Kamara, F.K., Tucker, V., Konuwa, E., Sellu, J., Mustapha, I., Foday, M., Yillah, M., Kanneh, F., Saffa*, S., Massally, J.L.B., Boisen, M.L., Branco, L.M., Vandi, M.A., Grant, D.S., Happi, C., Gevao, S.M., Fletcher, T.E., Fowler, R.A., Bausch, D.G., Sabeti, P.C., Khan*, S.H.,, Garry, R.F., for the KGH Lassa Fever Program, the Viral Hemorrhagic Fever Consortium, the WHO Clinical Response Team (* *deceased*), 2014. Clinical illness and outcomes in patients with Ebola in Sierra Leone. N. Engl. J. Med. 371 (22), 2092-2100.

Schmidt, M.W., Baldridge, K.K., Boatz, J.A., Elbert, S.T., Gordon, M.S., Jensen, J.H., Koseki, S., Matsunaga, N., Nguyen, K.A., Su, S.J., Windus, T.L., Dupuis, M., Montgomery, J.A., 1993. General atomic and molecular electronic-structure system. J. Comput. Chem. 14 (11), 1347-1363.

Science, 2014. The Science Contributors FAQ. URL http://www.sciencemag.org/site/feature/ contribinfo/faq/#pct_faq.

SCImago, 2014. Scientific Journal Rankings. URL http://www.scimagojr.com/journalrank.php? area=0&category=0&country=all&year=2012&order=sjr&min=0&min_type=cd.

Scopus, 2014. Database. URL http://www.elsevier.com/online-tools/scopus.

Scott, D.H., 1989. Secrets of Successful Writing, second ed. Reference Software International, San Francisco, CA.

Shamoon, H., Duffy, H., Fleischer, N., Engel, S., Saenger, P., Strelzyn, M., Litwak, M., Wylierosett, J., Farkash, A., Geiger, D., Engel, H., Fleischman, J., Pompi, D., Ginsberg, N., Glover, M., Brisman, M., Walker, E., Thomashunis, A., Gonzalez, J., Genuth, S., Brown, E., Dahms, W., Pugsley, P., Mayer, L., Kerr, D., Landau, B., Singerman, L., Rice, T., Novak, M., Smithbrewer, S., Mcconnell, J., Drotar, D., Woods, D., Katirgi, B., Litvene, M., Brown, C., Lusk, M., Campbell, R., Lackaye, M., Richardson, M., Levy, B., Chang, S., Heinheinemann, M., Barron, S., Astor, L., Lebeck, D., Brillon, D., Diamond, B., Vasilasdwoskin, A., Laurenzi, B., Foldi, N., Rubin, M., Flynn, T., Reppucci, V., Heise, C., Sanchez, A., Whitehouse, F., Kruger, D., Kahkonen, D., Fachnie, J., Fisk, J., Carey, J., Cox, M., Ahmad, B., Angus, E., Campbell, H., Fields, D., Croswell, M., Basha, K., Chung, P., Schoenherr, A., Mobley, M., Marchiori, K., Francis, J., Kelly, J., Etzwiler, D., Callahan, P., Hollander, P., Castle, G., Bergenstal, R., Spencer, M., Nelson, J., Bezecny, L., Roethke, C., Orban, M., Ulrich, C., Gill, L., Morgan, K., Laechelt, J., Taylor, F., Freking, D., Towey, A., Lieppman, M., Rakes, S., Mangum, J., Cooper, N., Upham, P., Jacobson, A., Crowell, S., Wolfsdorf, J., Beaser, R., Ganda, O., Rosenzweig, J., Stewart, C., Halford, B., Friedlander, E., Tarsy, D., Arrigg, P., Sharuk, G., Shah, S., Wu, G., Cavallerano, J., Poole, R., Silver, P., Cavicchi, R., Fleming, D., Marcus, J., Griffiths, C., Cappella, N., Nathan, D., Larkin, M., Godine, J., Lynch, J., Norman, D., Mckitrick, C., Haggen, C., Delahanty, L., Anderson, E., Lou, P., Taylor, C., Cros, D., Folino, K., Brink, S., Abbott, K., Sicotte, K., Service, F.J., Schmidt, A., Rizza, R., Zimmerman, B., Schwenk, W., Mortenson, J., Ziegler, G., Lucas, A., Hanson, N., Sellnow, S., Pach, J., Stein, D., Eickhoff, B., Woodwick, R., Tackmann, R., Trautmann, J., Rostvold, J., Link, T., Dyck, P., Daube, J., Colligan, R., Windebank, A., King, J., Colwell, J., Wood, D., Mayfield, R., Picket, J., Chitwood, M., 1993. The effect of intensive treatment of diabetes on the development and progression of long-term complications in insulin-dependent diabetes-mellitus. N. Engl. J. Med. 329 (14), 977-986.

Shekari, A., Hansen, K.M.S., Fankem, C., Tzakova, T., Patience, G.S., 2007. VPO Calcination: What are the Optimum Conditions? 20th NAM, Houston, TX.

Sheldrick, G.M., 1990. Phase annealing in SHELX-90: direct methods for larger structures. Acta Crystallogr. A 46 (6), 467-473.

Sheldrick, G.M., 2008. A short history of SHELX. Acta Crystallogr. A: Found. Crystallogr. 64 (1), 112-122.

Shepherd, A., Ivins, E.R., Geruo, A., Barletta, V.R., Bentley, M.J., Bettadpur, S., Briggs, K.H., Bromwich, D.H., Forsberg, R., Galin, N., Horwath, M., Jacobs, S., Joughin, I., King, M.A., Lenaerts, J.T.M., Li, J., Ligtenberg, S.R.M., Luckman, A., Luthcke, S.B., McMillan, M., Meister, R., Milne, G., Mouginot, J., Muir, A., Nicolas, J.P., Paden, J., Payne, A.J., Pritchard, H., Rignot, E., Rott, H., Sørensen, L.S., Scambos, T.A., Scheuchl, B., Schrama, E.J.O., Smith, B., Sundal, A.V., van Angelen, J.H., van de Berg, W.J., van den Broeke, M.R., Vaughan, D.G., Velicogna, I., Wahr, J., Whitehouse, P.L., Wingham, D.J., Yi, D., Young, D., Zwally, H.J., 2012. A reconciled estimate of ice-sheet mass balance. Science 338 (6111), 1183-1189.

Shewan, L.G., Coats, A.J.S., 2010. Ethics in the authorship and publishing of scientific articles. Int. J. Cardiol. 144, 1-2.

Shockley, W., Read Jr., W.T., 1952. Statistics of the recombination of holes and electrons. Phys. Rev. 87 (5), 835-842.

Silyn-Roberts, H., 2013. Writing for Science and Engineering Papers Presentations and Reports. Elsevier, Amsterdam, Netherlands.

Soares, J.B.P., Thomas, G., 2014. The Scientific Publishing Process: Important Tips on How to Write and Submit and Successful Paper, Short Course: 64th Canadian Chemical Engineering Conference, Niagara Falls.

Somvanshi, V.S., Sloup, R.E., Crawford, J.M., Martin, A.R., Heidt, A.J., Kim, K., Clardy, J., Ciche, T.A., 2012. A single promoter inversion switches *Photorhabdus* between pathogenic and mutualistic states. Science 337 (6090), 88-93.

Spek, A.L., 2003. Single-crystal structure validation with the program PLATON. J. Appl. Crystallogr. 36 (1), 7-13.

Sprunck, S., Rademacher, S., Vogler, F., Gheyselinck, J., Grossniklaus, U., Dresselhaus, T., 2012. Egg cell-secreted EC1 triggers sperm cell activation during double fertilization. Science 338 (6110), 1093-1097.

Starink, M.J., 2003. The determination of activation energy from linear heating rate experiments: a comparison of the accuracy of isoconversion methods. Thermochim. Acta 404 (1-2), 163-176.

Steyerberg, E.W., Vickers, A.J., Cook, N.R., Gerds, T., Gonen, M., Obuchowski, N., Pencina, M.J., Kattan, M.W., 2010. Assessing the performance of prediction models: a framework for some traditional and novel measures. Epidemiology 21 (1), 128-138.

Strunk Jr., W., White, E.B., 2000. The Elements of Style, fourth ed. Pearson Education, Inc., Upper Saddle River, NJ.

Sun, L., Wan, S., Luo, W., 2013. Biochars prepared from anaerobic digestion residue, palm bark, and eucalyptus for adsorption of cationic methylene blue dye: characterization, equilibrium, and kinetic studies. Bioresour. Technol. 140, 406-413.

Suntivich, J., May, K.J., Gasteiger, H.A., Goodenough, J.B., Shao-Horn, Y., 2011. A perovskite oxide optimized for oxygen evolution catalysis from molecular orbital principles. Science 334 (6061), 1383-1385.

Swietoslawski, M., Swider, J., Molenda, M., 2014. Novel Co-precipitation Method of LiFePO$_4$ Synthesis in an Anhydrous Environment. OREBA, Montreal.

Sword, H., 2012, 23 July. Zombie Nouns. New York Times, URL http://opinionator.blogs.nytimes.com/2012/07/23/zombie-nouns/?_r=0.

Tabkhi, F., Pibouleau, L., Hernandez-Rodriguez, G., Azzaro-Pantel, C., Domenech, S., 2010. Improving the performance of natural gas pipeline networks fuel consumption minimization problems. AIChE J. 56 (4), 946-964.

Tamura, K., Dudley, J., Nei, M., Kumar, S., 2007. MEGA4: molecular evolutionary genetics analysis (MEGA) software version 4.0. Mol. Biol. Evol. 24 (8), 1596-1599.

Tamura, K., Peterson, D., Peterson, N., Stecher, G., Nei, M., Kumar, S., 2011. MEGA5: molecular evolutionary genetics analysis using maximum likelihood, evolutionary distance, and maximum parsimony methods. Mol. Biol. Evol. 28 (10), 2731-2739.

Tarduno, J.A., Cottrell, R.D., Nimmo, F., Hopkins, J., Voronov, J., Erickson, A., Blackman, E., Scott, E.R.D., McKinley, R., 2012. Evidence for a dynamo in the main group pallasite parent body. Science 338 (6109), 939-942.

Tavares, J.R., Virgilio, N., 2014. Oral Presentation Techniques, Workshop for the Chemical Engineering Students, 2014 Polytechnique Montréal.

Taylor, M., 2013. Plagiarism is nothing to do with copyright. URL http://svpow.com/2013/09/20/plagiarism-is-nothing-to-do-with-copyright/.

Terranova, O., Antronico, L., Coscarelli, R., Iaquinta, P., 2009. Soil erosion risk scenarios in the Mediterranean environment using RUSLE and GIS: an application model for Calabria (southern Italy). Geomorphology 112 (3-4), 228-245.

Thomas, M.K., Kremer, C.T., Klausmeier, C.A., Litchman, E., 2012. A global pattern of thermal adaptation in marine phytoplankton. Science 338 (6110), 1085-1088.

Thompson, A., Taylor, B.N., 2008. Guide for the Use of the International System of Units (SI): NIST Special Publication 811 2008 Edition. National Institute of Standards and Technology. URL http://physics.nist.gov/cuu/pdf/sp811.pdf.

Thompson, J.D., Gibson, T.J., Plewniak, F., Jeanmougin, F., Higgins, D.G., 1997. The CLUSTAL_X windows interface: flexible strategies for multiple sequence alignment aided by quality analysis tools. Nucleic Acids Res. 25 (24), 4876-4882.

Thompson, J.D., Higgins, D.G., Gibson, T.J., 1994. CLUSTAL W: improving the sensitivity of progressive multiple sequence alignment through sequence weighting, position-specific gap penalties and weight matrix choice. Nucleic Acids Res. 22 (22), 4673-4680.

Thring, R.W., Chornet, E., Overend, R.P., 1990. Recovery of a solvolytic lignin: effects of spent liquor acid volume ratio, acid concentration and temperature. Biomass 23 (4), 289-305.

Tu, W., Chin, Y.H.C., 2014. A Generalized Thermodynamic Activity Relation of Metal and Oxide Clusters and Their Periodic Reactivity Trend for Methanol Oxidative Dehydrogenation, 2014 AICHE Annual Meeting, Memphis, TN.

Tuck, C.O., Pérez, E., Horváth, I., Sheldon, R.A., Poliakoff, M., 2012. Valorization of biomass: deriving more value from waste. Science 337 (6095), 695-699.

University of Toronto, 2014. Plagiarism: Definitions, Examples and Penalties Department of Chemistry, University of Toronto. URL http://www.chem.utoronto.ca/undergrad/plagiarism.php.

Valadés-Pelayo, P.J., Sosa, F.G., Serrano, B., de Lasa, H., 2015. Photocatalytic reactor under different external irradiance conditions: validation of a fully predictive radiation absorption model. Chem. Eng. Sci. 126, 42-54.

van Loo, A.F., Fedorov, A., Lalumière, K., Sanders, B.C., Blais, A., Wallraff, A., 2013. Photon-mediated interactions between distant artificial atoms. Science 342 (6165), 1494-1496.

Vanderbilt, D., 1990. Soft self-consistent pseudopotentials in a generalized eigenvalue formalism. Phys. Rev. B 41 (11), 7892-7895.

Vieira, E.S., Gomes, J.A.N.F., 2010. Citations to scientific articles: its distribution and dependence on the article features. J. Informetrics 4 (1), 1-13.

Voelker, R., 2014. Regulation boosts vaccination rates. J. Am. Med. Assoc. 311 (18), 1848.

Vyazovkin, S., Burnham, A.K., Criado, J.M., Pérez-Maqueda, L.A., Popescu, C., Sbirrazzuoli, N., 2011. ICTAC Kinetics Committee recommendations for performing kinetic computations on thermal analysis data. Thermochim. Acta 520 (1-2), 1-19.

Wang, D.X., Chen, Y.G., Wei, S., Lia, X.M., 2014a. Synthesis, crystal structure and properties of the bichromium(III)-substituted tungstosilicate: $[N(CH_3)_4]_4$ $[A-\beta_{(12)}-SiW_{10}Cr_2O_{36}(OH)_2(H_2O)_2] \cdot 5H_2O$. Inorg. Chim. Acta 415 (1), 146-150.

Wang, X., Li, Z., Shi, J., Yu, Y., 2014b. One-dimensional titanium dioxide nanomaterials: nanowires, nanorods, and nanobelts. Chem. Rev. 114, 9346-9384.

Wang, H., Qiao, X., Chen, J., Ding, S., 2005. Preparation of silver nanoparticles by chemical reduction method. Colloids Surf. A: Physicochem. Eng. Asp. 256 (2-3), 111-115.

Wang, T., Chen, J., Yang, T., Xiao, C., Sun, Z., Huang, L., Dai, D., Yang, X., Zhang, D.H., 2013. Dynamical resonances accessible only by reagent vibrational excitation in the F + HD → HF + D reaction. Science 342 (6165), 1499-1502.

Ware, J.E., Sherbourne, C.D., 1992. The MOS 36-item short-form health survey (SF-36). 1. Conceptual-framework and item selection. Med. Care 30 (6), 473-483.

Weaver, D.F., Barden, C., 2009. Don't overlook the rigorously reviewed novel work in patents. Nature 461.

Web of Science™, 2014. Thomson Reuters Web of Science. URL apps.webofknowledge.com/ WOS/_ClearGeneralSearch.do?action=clear&product=WOS&search/mode=GeneralSearch& SID.

Weedon, G.P., Gomes, S., Viterbo, P., Shuttleworth, W.J., Blyth, E., Österle, H., Adam, J.C., Bellouin, N., Boucher, O., Best, M., 2011. Creation of the WATCH forcing data and its use to assess global and regional reference crop evaporation over land during the twentieth century. J. Hydrometeorol. 12 (5), 823-848.

Whitesides, G.M., 2004. Whitesides' group: writing a paper. Adv. Mater. 16 (15), 1375-1377.

Winzer, T., Gazda, V., He, Z., Kaminski, F., Kern, M., Larson, T.R., Li, Y., Meade, F., Teodor, R., Vaistij, F.E., Walker, C., Browser, T.A., Graham, I.A., 2012. A *Papaver somniferum* 10-gene cluster for synthesis of the anticancer alkaloid noscapine. Science 336, 1704-1708.

Wolke, J.G.C., Dijk, K.V., Groot, H.G.S.K.D., Jansen, J.A., 1994. Study of the surface characteristics of magnetron-sputter calcium phosphate coatings. J. Biomed. Mater. Res. 28 (12), 1477-1484.

Wolpaw, J.R., Birbaumer, N., McFarland, D.J., Pfurtscheller, G., Vaughan, T.M., 2002. Brain-computer interfaces for communication and control. Clin. Neurophysiol. 113 (6), 767-791.

Won, J.Y., Jun, H.K., Jeon, M.K., Woo, S.I., 2006. Performance of microchannel reactor combined with combustor for methanol steam reforming. Catal. Today 111 (3-4), 158-163.

Yan, B., Cheng, Y., Jin, Y., 2013. Cross-scale modeling and simulation of coal pyrolysis to acetylene in hydrogen plasma reactors. AIChE J. 59 (6), 2119-2133.

Yang, H., Heo, J., Park, S., Song, H.J., Seo, D.H., Byun, K.E., Kim, P., Yoo, I., Chung, H.J., Kim, K., 2012. Graphene barristor, a triode device with a gate-controlled Schottky barrier. Science 336 (6085), 1140-1143.

Yang, T.T., Cheng, L.Z., Kain, S.R., 1996. Optimized codon usage and chromophore mutations provide enhanced sensitivity with the green fluorescent protein. Nucleic Acids Res. 24 (22), 4592-4593.

Yoffe, A.D., 2001. Semiconductor quantum dots and related systems: electronic, optical, luminescence and related properties of low dimensional systems. Adv. Phys. 50 (1), 1-208.

Yoshida, T., Maruyama, T., Akizuki, Y., Kan, R., Kiyota, N., Ikenishi, K., Itou, S., Watahiki, T., Okano, H., 2013. SPARC64 X: Fujitsu's new-generation 16-core processor for UNIX servers. IEEE Micro 33 (6), 16-24.

Zappoli, B., 1990. Response of a solid-gas growth interface to a homogeneous time dependent acceleration field. Int. J. Heat Mass Transf. 33 (9), 1829-1837.

Zhang, Q., Hwang, J.W., Kim, K.N., Jung, H.W., Noh, S.M., Oh, J.K., 2013a. New photo-induced thiol-ene crosslinked films based on linear methacrylate copolymer polythiols. J. Polym. Sci. A: Polym. Chem. 51 (13), 2860-2868.

Zhang, W., Oganov, A.R., Goncharov, A.F., Zhu, Q., Boulfelfel, S.E., Lyakhov, A.O., Stavrou, E., Somayazulu, M., Prakapenka, V.B., Konôpková, Z., 2013b. Unexpected stable stoichiometries of sodium chlorides. Science 342 (6165), 1502-1505.

Zhao, S.-q., Liu, G., Ou, Q.-h., Xu, J., Ren, J., Hao, J.-m., 2014. Analysis of different types of soil by FTIR and ICP-MS. Spectrosc. Spectr. Anal. 34 (12), 3401-3405.

Zhou, H., Yang, L., Stuart, A.C., Price, S.C., Liu, S., You, W., 2011. Development of fluorinated benzothiadiazole as a structural unit for a polymer solar cell of 7 % efficiency. Angew. Chem. 123 (13), 3051-3054.

Zielinska, E., 2011. Poster perfect—how to drive home your science with a visually pleasing poster. URL http://www.the-scientist.com/?articles.view/articleNo/31071/title/Poster-Perfect/.

Index

Note: Page numbers followed by *f* indicate figures and *t* indicate tables.